MULTINATIONALS, MINING AND DEVELOPMENT

To T.T. and the memory of C.K.T.

# Multinationals, Mining and Development

## A Study of the Tin Industry

JOHN THOBURN
*University of East Anglia, Norwich*

Gower

Published by
Gower Publishing Company Limited
Westmead, Farnborough, Hampshire, England

British Library Cataloguing in Publication Data

Thoburn, John
Multinationals, mining and development.
1. Tin mines and mining
2. Tin industry
3. Corporations, International
I. Title
338.8'8 HD9539.T6

ISBN 0-566-00417-8

PRINTED BY NENE LITHO
BOUND BY WEATHERBY WOOLNOUGH

# Contents

# Preface

This study has been written over the period April 1978 to September 1980 with the aid of a grant from the Overseas Development Administration of the Foreign and Commonwealth Office, London, for which I am most grateful. ODA provided research expenses throughout, including those of two visits to South East Asia, and salary during the summer term 1979 which I spent as a Visiting Fellow at the Institute of Development Studies in the University of Sussex. I have also benefitted from an opportunity to work as a consultant to the UN Economic and Social Commission for Asia and the Pacific on transnational corporations in the South East Asian tin industry.

The project has been associated with a wider range of commodity studies organised from IDS, Sussex. The commodities group at Sussex was formed in the belief that research over a number of mineral and agricultural products exported from less developed countries would yield information of value to policy makers in increasing the gains from commodity exports, on which the Third World still depends for over three quarters of its export earnings. The group has proved a stimulating forum in which researchers of often quite widely divergent political and economic views could exchange ideas; and I should like to single out for thanks in this regard David Evans and Carlos Fortin of IDS, Bill Page from the Science Policy Research Unit at Sussex, and Chris Edwards of UEA. Warm thanks are also due to Hans Singer, from whom came the invitation to visit IDS, and who consistently has provided kindly and helpful support to my research in numerous ways.

During visits to Malaysia and Thailand in 1978 and Indonesia in 1979 I received considerable assistance from the Mines Department in Kuala Lumpur, the Department of Mineral Resources in Bangkok and from P. T. Tambang Timah and the Department of Mines and Energy in Jakarta. I should like to thank the officials of these agencies and, in addition, the many mining companies who provided information. Umar Yahya and Bernard Engel of the International Tin Council in London also have taken a helpful interest in the research.

In a comparative study of several countries material in different local languages can present a problem. Suchart Thamrongrech and Soonthorn Vongurai have translated several works from Thai. Similar help has been provided by Sritua Arief with material in Indonesian and Elizabeth Onians has translated work on Indonesia from Dutch. David Lee of the University of Essex and Alan Camina of UEA helped with some mathematical points, and statistical assistance was provided by Jacquie White. The index was prepared by Liz Edwards, camera-ready copy by Mary Gurteen, and earlier drafts were typed by Ros Cruickshank, Christine Jope and Pauline Speculand. I thank them all.

Of course, neither ODA nor any of the persons or organisations mentioned here necessarily share any view expressed in the study, and I alone am responsible for any errors.

JTT

Norwich November 1980

# Abbreviations

| | |
|---|---|
| AR | Annual Report |
| BSMI | Bulletin of Statistics Relating to the Mining Industry (Malaysia) |
| DMR | Department of Mineral Resources, Bangkok |
| ESCAP | Economic and Social Commission for Asia and the Pacific, United Nations, Bangkok |
| FEER | Far Eastern Economic Review |
| ILO | International Labour Office, Geneva |
| ITA | International Tin Agreement |
| ITC | International Tin Council, London |
| LDC | Less Developed Country |
| MMC | Malaysia Mining Corporation |
| MNC | Multinational Company (or Corporation) |
| MSB | Monthly Statistical Bulletin |
| OMO | Offshore Mining Organisation, Thailand |
| TI | Tin International, London |
| TNC | Transnational Company (or Corporation) |
| UNCTAD | United Nations Conference on Trade and Development |
| UNDP | United Nations Development Programme |
| UNIDO | United Nations Industrial Development Organisation |

# 1 Introduction

This study is concerned with the methods by which less developed countries can increase the contribution of mineral exports to their socio-economic development.(1) Following the effective cartelisation of the oil industry in 1973-74 much stress has been laid in the United Nations Conference on Trade and Development and other Third World forums on increasing commodity prices. Yet in the past many potential gains have been lost to the exporting countries through import leakages, tax avoidance through transfer pricing, and failure to process the product locally; and excessively capital-intensive technology has led to adverse effects on local income distribution and the level of employment. It is not contended here that the level of commodity prices (or their stability) is unimportant, but primary stress is laid on the restructuring of production and marketing.(2)

Restructuring issues are closely connected with questions of technology and ownership. The choice of technique of production is probably the most important single influence on an export's development effects, conditioning employment generation, distribution of value-added between wages and profits, intermediate purchases and sales, and so on; and many of these effects in turn determine the spread of economic activity beyond the original export sector and thus help generate a base for independent longer term development. This independence is of vital importance in the case of mineral exports, where ultimately the resource will be depleted. Ownership not only gives claim to profits but also allows control of the choice of technique. Foreign ownership is thought often to have led to the use of more capital-intensive methods than would seem appropriate in the light of the host countries' prevailing resource endowment, and indeed foreign firms historically in some cases have developed capital-intensive technology where existing techniques were labour-intensive. Moreover, ownership gives control over input purchases and the location of product processing; and international companies, especially if vertically integrated, may be able to exercise considerable market power *vis-à-vis* host economies by switching their sources of supply or investing profits from one country in expansion in another; or by controlling marketing even if the production of the commodity is under local control.

In minerals, foreign ownership is of particular importance since in recent years the industry has grown extremely capital-intensive. Projects of half a billion or even a billion dollars are not unusual and even LDC governments find difficulty in raising sums of this kind, let alone local LDC companies. Allied with these increased capital requirements is a significant growth in the complexity of technology, especially as computerised methods are used to operate the mine. In most minerals there has been an accompanying tendency to switch from underground operations to larger, low grade open cast ones. The very size, complexity, and high cost of new mining ventures frequently lead international lending agencies and commercial lenders to require the participation of multinational mining companies since the lenders feel that only such companies have the necessary expertise. Nevertheless there has also been a growth in state ownership of mining in LDC's(3) often as a result of the takeover of facilities originally owned by MNC's.(4) Private local ownership of the mines is still sometimes an alternative to foreign direct investment, however.

Policy measures to increase host-country gains have included taxes on profits or export revenue to channel income back into the domestic economy or to influence the rate of resource depletion, tax incentives encouraging, or regulations requiring, higher local content of intermediate products, more labour-intensive methods, or more local processing. Outright nationalisation has also been tried by several mineral exporting countries, as has the compulsory restructuring of equity. Many problems face such countries, however, in formulating and implementing policy. Available labour-intensive methods may not be efficient even if inputs are revalued at shadow prices, heavy taxation may discourage foreign investment, and production technology may be monopolised by foreign companies, while local firms may be unable to develop it themselves. Local processing of output or local production of some intermediate products also may not always be profitable privately or socially. In the face of such problems studies of individual minerals over a range of producing countries may help policy makers learn from the experience of other countries and suggest hitherto untried possibilities. Investigating the structure of ownership and control where foreign companies are involved will, by identifying horizontal links between producing countries and vertical links backwards to equipment suppliers or forwards to shippers, marketers, fabricators, etc., help assess the market power exertable by the foreign firms and ultimately help indicate whether continued foreign investment is desirable.

Tin is unusual in that, until the 1960's at least, there was very little participation in the industry by the major mining multinationals, and tin companies were specialised. This is striking since most mining companies spread into a large number of metals or are highly integrated vertically, or both.(5) Three other features are of importance in tin. First, there coexists a range of production techniques in most producing countries, and the more labour-intensive techniques have been associated with continued local ownership. Second, there are great differences in costs between different groups of producers, generating large mineral rents in the intramarginal countries. Bargaining over these rents leads to a special interest in tax policy and in measures to alter the structure of share-ownership. Third, the industry is highly polarised, in the sense that most tin exports come from the Third World while most tin is consumed in developed countries. This inevitably leads to interest in the industry's cartelisation potential as a supplement or alternative to restructuring.

This study then is concerned mainly with the changing structure of the tin industry in terms of foreign investment and its alternatives; with the development gains which have accrued to host-countries through the operation of market forces; and with attempts by governments to shift the distribution of benefits from tin production by foreign companies in favour of the host-country. The project sees itself as integrative, drawing together material from existing studies and filling in gaps by new work where necessary. It concentrates on the four largest producers - Malaysia, Bolivia, Thailand and Indonesia - who produce about two-thirds of world output, though minor producers also are discussed quite frequently. Much more new work is presented on the three South East Asian countries than on Bolivia, since a Harvard Institute of International Development team reported extensively on Bolivia's tin industry in the mid 1970's.(6).

The study opens by discussing (in chapter two) mineral development policy in the context of theories of the gains from mineral exports and foreign investment. Chapter three gives an overview of the structure of the tin industry, including its technology and organisation and describes the individual tin exporting countries. It is shown that the main exporting countries are all committed, at least in theory, to policies of income redistribution as well as growth, and mining policy has to be considered in the light of these expressed priorities.(7) Chapter four traces the involvement of foreign companies in the industry, and their interaction with local mining firms, and chapter five looks at recent changes in policy. In the tin industry there has been a wide range of alternative policies pursued, from nationalisation in Bolivia and Indonesia to a stock exchange takeover in Malaysia, while Thailand has seen the almost explosive growth of a new local mining sector in the form of suction boats. Chapter five also attempts to quantify the gains from existing local production. Chapter six, which has occupied a large part of the project's research time, considers the effects of present policies on new investment in South East Asia. It asks whether the present tax levels are so high as to discourage investment, and whether the tax structures need changing (particularly with regard to the mix between taxes on output and taxes on profit). Using social cost-benefit techniques, it investigates the economic and social profitability of alternative techniques of production and evaluates different degrees of state equity participation in new mining ventures. Chapter seven gives a summary and some policy recommendations.

NOTES

(1) Many of the conclusions of the study will be relevant to agricultural exports too. Minerals have the added dimensions that their stocks are exhaustible (so that rates of depletion must be considered) and that they often generate substantial economic rents (hence special attention is given in mineral studies to taxation and local equity participation).
(2) This interest in arrangements for commodity trade and production which are wider than the traditional UNCTAD objectives of 'stable and remunerative' prices is also now part of official United Nations policy, as expressed in the UN demands for a New International Economic Order and in UNCTAD's Integrated Programme for Commodities. For a far-reaching discussion of the UN's commodity objectives see Evans (1979). For a very useful introductory account of the 'structural' approach to commodities see Singer (1976).
(3) For general discussion of these issues see Bosson and Varon (1977) and Mikesell (1979).
(4) The terms 'multinational', 'transnational' and 'international' company are used here interchangeably. It is not intended that 'MNC' should denote a wider spread of countries or activities than 'TNC'. The terms simply denote companies operating outside their country of

origin. This follows the now common usage, but there remains a problem in that the older, specialist tin companies, though they are MNC's, are much less widely spread than some of the very large new entrants to the tin industry, who are already active in many other metals. Such highly diversified companies usually will be referred to here as major mining multinationals.

(5) See Lanning and Mueller (1979) for an interesting (and provocative) discussion of international mining companies.

(6) The report has been published as a book (Gillis, 1978). A HIID team also reported on Indonesia's mining industry in 1977, but its report is not available for quotation.

(7) The commitment of the new Bolivian military junta to these ideals is not clear, but the previous governments have been committed to them.

# 2 The Development Gains from Mineral Exporting and Foreign Investment

This chapter is concerned with the identification and the methods of measurement of potential further gains from mineral exports. In view of the widespread involvement of foreign companies in LDC mineral production mentioned in the previous chapter, and their changing importance in tin, this is tantamount to an analysis of the benefit to the host-country of foreign direct investment and its alternatives. During the last decade a substantial body of work has accumulated on mineral development policy, especially with regard to depletion rates and resource rents, and to tax policy in relation to both of these. There is also much useful information to be gleaned from the mainstream development literature. Thus in the 1950's and early 1960's there was an outpouring of work on trade and development, which is covered in section 1. Most of this was concerned with the efforts of primary commodity exporting and its merits *vis-à-vis* manufacturing exports or import substitution, and it contains various comparisons of the development effects of mining and agricultural exports. The issues which this highlights are those of technology (with its influence on linkages, income distribution, employment creation and the spreading of labour skills) and foreign ownership (which influences the choice of technique and local purchasing policy, and may affect domestic savings and the supply of local entrepreneurship). Since then most work has become more specialised with literatures growing up on appropriate technology, on poverty, employment and income distribution, and on foreign investment and multinational companies. This is sifted in section 2; and the first two sections also consider briefly the negative effects of exports and foreign investment from the perspective of the 'underdevelop -ment' school. Section 3 then looks at how the benefits from foreign investment and its alternatives can be measured. Section 4 discusses mineral development policy, with special reference to taxation.

## 2.1 TRADE, DEVELOPMENT AND UNDERDEVELOPMENT (1)

Classical economists, especially Smith and Mill, had held an extremely optimistic view of the effects of foreign trade on growth. Basically, their position was founded on an unbalanced growth argument. Export expansion, by definition, was not constrained by domestic market size. Beyond the comparative advantage, reallocative gains from trade (of which Smith's 'vent for surplus' was the extreme case), such expansion would stimulate new capital formation and technical change. Smith stressed the scope for improved division of labour, and Mill the opportunities for more mechanised production. Much later writers such as Stigler (1951) have shown how real costs could be lowered by specialisation both within and between firms in this way. Marshall too believed in the advantages of 'massive production' and that these could be gained from trade. Imports from developed countries into poor ones were, for Mill, a way of stimulating new tastes and economic aspirations, and the higher profits obtainable from newly founded export industries were an inducement to an increased propensity to save. Apart from criticism by

Marxist writers (about which more later) the Classical position remained largely unchallenged until after the Second World War.

Most of the work on trade and growth in the 1950's and early 1960's had its origins in the writings of Singer, Myrdal, and Prebisch. (2) The 'SMP' position can be summarised quite briefly. From Singer, who was explicitly concerned with foreign ownership, derives the notion of the export sector as an enclave, located in the host-country only in a geographical sense while economically part of the investing country. Myrdal, concerned with primary production versus manufacturing, adds the ideas of 'spread' and 'backwash' effects. The former, by which growth could be spread through the economy by demands for consumer goods, agricultural products, etc., are said to be weak because of the unintegrated nature of a less developed country. 'Backwash' effects, however, may be strong and they include the stifling of local industry by imported western products, and the drawing away from the rest of the economy of its best productive resources of capital and entrepreneurship. Since the bulk of export production and foreign investment at the time was in primary commodities,(3) this drawing off of resources would retard local manufacturing development. In general, though it was not denied that primary export production would lead to some of the cumulative additions to income and technical knowledge envisaged by the Classical economists, these were contended to be greater for manufactures. Hence the stifling of local manufacturing was seen as particularly harmful and many subsequent arguments for import substituting industrialisation were based on the SMP position.

Such gains as there were from primary production did not accrue to the host-country, it was argued. There were two main channels through which the gains could be lost. First, there were leakages in the form of remitted profits, imported consumer goods (especially for foreign skilled labour), and imported intermediate and capital goods; in addition to which the primary products tended to be exported in a minimally processed form. The other channel was the alleged secular deterioration of the (barter) terms of trade for primary producers(4) a point which was strongly argued by Prebisch and the UN Economic Commission for Latin America. This is brought about by the low income elasticity of demand for primary products in relation to manufactures; by the apparent inability of LDC workers (and capitalists) to translate productivity increases into higher wages (or profits) rather than lower prices; and the rigidity of resource use such that in the face of slowly rising demand and increasing productivity, price falls could not be forestalled by the reallocation of resources to alternative uses.(5) These two channels are mutually exclusive to a large extent, though the inconsistency or arguing for both at once is rarely noted. There is little sense in foreign investment in commodity exports, the gains from which are being transferred to western consumers in the form of falling relative prices, unless the investors are themselves consumers (metal fabricating firms opening mines, etc.) In the latter case the exports would probably be traded at transfer prices anyway and the free-market price (because it clears only residual supplies) would not be a reliable guide to market conditions. Conversely, if there are gains to the investors from primary production per se, this is as good an argument for local ownership (or greater government control) of the activity as it is for diversification into manufacturing. There is, however, a possibility that local producers may be exploited by the foreign investors, e.g. where peasants sell their output to foreign estates for marketing, or small miners sell ore to a foreign owned local smelter.

Naturally, the SMP position provoked much controversy. Bauer (1959) argued that enclaves were a focal point of development. Hicks (1959) stressed how trade at least allowed goods or low growth potential (e.g. primary commodities) to be exchanged for goods of high growth potential (e.g. capital goods). Myint (e.g. 1954 and 1967) provided a quite interesting explanation of limited economic development, though mainly with regard to agricultural exports, in terms of the way in which LDC's were opened to trade. In the case of peasant exports, the key feature was that little change in productive technique was required, the export expansion being a use of otherwise surplus land and labour. Thus the gain in income was of a once-and-for-all kind, though presumably profits could be reinvested for further growth. In the case of plantation exports a cheap labour policy became fossilised, often with the importation of workers from low wage areas (such as the Indian labour used in the rubber estates of Malaya or the tea estates of Ceylon), which of course increased the isolation of the sector from the rest of the economy.

The SMP thesis, though full of interesting ideas and empirical assertions, was lacking in a well thought out analysis of the mechanisms through which exports might lead to growth. Such an analysis was to come from the work of Baldwin (1956 and 1963) and Watkins (1963) drawing heavily on the 'staple' approach developed by economic historians to explain the early growth of Canada and of regions in the United States. A 'staple' is any natural resource-intensive export, and one task of any theory of exports and development clearly is to explain how some countries such as Canada and Australia seem to have achieved high <u>per capita</u> income levels through specialisation on commodity exports, while many LDC's have not. The key to the 'staple' approach is the finding that different export products in a given environment have different effects on development, and that these differences are not simply associated with a 'primary products <u>versus</u> manufactures' distinction. Thus the Canadian historian Harold Innis showed how fur exports had retarded Canadian development whereas exports encouraged it. (Innis, 1930 and 1956)

Essentially, the staple theory (which is really the high point of the 'trade and development' literature) is one of trade induced capital formation and technical change. An export opportunity, perhaps initiated by foreign investment, may stimulate domestic firms to enter the industry. Opportunities to produce intermediate products or capital goods for the export sector, to use the export good as an input for further processing, or to produce consumer goods for workers in the sector, likewise may stimulate investment. These opportunities are, of course, <u>linkages</u> (6) in Hirschman's (1958) terminology. If the country is short of investment outlets, linkages may raise the overall level of net investment and stimulate new saving. Even if the linkages are simply reallocations of existing investment resources, new technology may be introduced with them which raises the productivity of local capital and labour. In addition, workers may improve their skills through these new employments, and managers and entrepreneurs may learn new and useful methods. These improvements may, in turn, increase the responsiveness of the domestic economy to linkage and other new investment opportunities, for which export sector profits may help provide capital. Linkage investment too (e.g. in transport) may stimulate growth in other sectors.

But in such terms the staple theory appears both highly optimistic and very neoclassical. In fact, it has considerable potential to explain the lack of development in some circumstances. The analysis stresses the disposition of export income flows, and is organised around the export product's production function and potential input-output relations. Foreign investment may use a technology which precludes local entry into export production, either by its sheer complexity or its high initial capital requirements. Similar conditions may apply to backward linkages, whose products may either be imported or produced locally by foreign firms, and forward linkage products may be produced overseas.(7) The less vertically integrated the export industry can be, the smaller is the initial investment, and therefore the greater are the opportunities for linkage investment and diversification away from the export base. Diversification of this kind is especially important where minerals are concerned, since a foundation must ultimately be created for self-sustained development once the resource is exhausted. However, as Little and Mirlees (1974) have pointed out, backward linkages could develop on their own through direct export; and forward linkages similarly could import their inputs Hirschman's point, though, is that such opportunities may be difficult to perceive without the initial activity coming into being.

Besides the expenditure on intermediate products (and capital goods), the split in value added between wages and profits is of great importance, even where the export activity is locally owned. The share of wages (and the composition of the wages bill), as well as having a direct influence on income distribution and employment creation, is a major factor in local market development.(8) These differences are well illustrated by Baldwin's (1963) characterisation of mineral and plantation exports. Minerals, he argues, typically employ small numbers of skilled workers, often expatriates, who tend to demand luxury consumer goods which are often imported. Plantations normally employ very large unskilled labour forces at low wages. The plantation labour is therefore likely to generate a market for local consumer goods, but they will be of a very simple kind (often little more than basic foodstuffs)which will require few changes in techniques. With regard to backward linkages, minerals are likely to require sophisticated capital goods and intermediate products, probably imported, whereas plantations required few purchased intermediate products. Local entry is said to be precluded in minerals because of the advanced nature of the technology, and plantations require specialised and costly management which has a similar effect. In an earlier paper Baldwin (1956) compares the plantation crop with the archetypal family crops which led to growth in Canada and the early United States of America. The family farm crop required moderately skilled labour, was far more labour-intensive than most minerals, and the egalitarian distribution of income associated with it gave rise to a substantial and diversified local market for consumer goods. Also, the family farm crop was associated with a particularly advantageous kind of factor inflow -

fairly well educated workers, often with some capital of their own. In contrast, plantations frequently imported very cheap labour which was often isolated socially and economically from the host economy. Whether the tin mining industry in LDC's corresponds to these archetypes, and whether small scale tin mining and family farm crops have much in common will emerge later in this study.(9) What is worth noting at this stage is that Baldwin's archetypes are essentially explanations of 'underdevelopment'.

The Singer-Myrdal-Prebisch position already summarised in this section clearly established the possibility that the development effects of commodity exports, especially if undertaken by foreign investors, might be slight or even negative. Negative development effects - or trade and foreign investment as forces to 'underdevelop' poor countries - have been fitted into a systematic framework as part of various 'dependency' viewpoints and have also been considered extensively by Marxist writers working on imperialism. Dependency theory also fits well into the appropriate technology debate set out in the next section.

Although much recent 'dependency' work is concerned with agricultural and mineral exports (and mainly those controlled by foreign investors) the origins of dependency writings are in reactions to the results of import-substituting industrialisation in Latin America - ironically the very kind of policies proposed by Prebisch and ECLA. They describe the lopsided pattern of development brought about by excessively capital-intensive technology (see section 2) but beyond that (and merging with some Marxist work(10)) they show how different forms of production (e.g. for internal élites) may condition the formation and continuation of class structures, and indeed bring about the reproduction of dependency. Frank (1966), moreover, has suggested a vertical chain of exploitation, whereby a series of economic and social relations transfer resources from satellite to metropolitan areas within LDC's, and thence to rich countries. The Caribbean dependency writers such as Girvan (1970) looking at mineral exports have also provided a number of specific and testable mechanisms. These include the loss of linkages to overseas branches of multinational companies, but the stress is on lopsided growth and unemployment of the kind described in the next section. The notion of secularly declining terms of trade has been formalised in recent writings by Emmanuel and others on unequal exchange.(11)

The Marxist theory of imperialism, where firms invest overseas to extract surplus value from cheaper overseas labour and thereby forestall the fall in their rate of profit, of course, long antedates Singer-Myrdal-Prebisch. Barratt-Brown (1974, p.71) however argues that until 1928 (the 1928 Congress of the Comintern) Marxists believed that imperialism transmitted development to LDC's, albeit development of a capitalist kind with all its inherent contradictions. After that date the official view was that imperialism acted as a force for underdevelopment, under which LDC workers and capitalists get exploited, though there is conceded to be a 'compradox' bourgeosie in LDC's which gains from serving the interests of the imperialist countries.

Finally it is worth noting that the staple theory itself has been reconsidered by Canadian writers working from a Marxist perspective, and they argue that concentration on staple exports has tended to 'underdevelop' Canada.(12) While the notion that Canada is 'dependent' (mainly on the USA) and structurally underdeveloped seems (to this writer at least) to be of little relevance, given its very high per capita income, one substantive point is made which is that international mining companies tend not to reinvest their profits in anything other than more mining and that they will invest abroad if further local mining investment opportunities are lacking.

## 2.2 FOREIGN INVESTMENT AND APPROPRIATE TECHNOLOGY

It is now well established that the technology in use in LDC's often is 'inappropriate' in the sense that high capital/labour ratios are used in situations of relative labour abundance. In some cases this may reflect the fact that no efficient labour-intensive processes exist. This is where the capital-intensive methods use less capital per unit of output as well as less labour (i.e. the labour-intensive methods are 'dominated' by the capital-intensive ones). Alternatively, firms may choose capital-intensive production methods because factor prices are distorted, with the result that the techniques are privately but not socially the most profitable. Further, firms may even act in apparent defiance of factor prices, reflecting that they have a wider set of objectives by choosing capital-intensive techniques which seem neither privately nor socially the most profitable. In practice too, foreign firms (and also large local ones) using high capital intensity may face factor prices which differ from those faced by smaller firms. However, for foreign firms as a whole, this is as likely to be a consequence as an initial cause of high capital-intensity.

Stewart (1978 pp. 116) has depicted a situation in which the use of capital-intensive technology becomes self-perpetuating. Techniques are developed in western countries to suit western conditions and trans-ferred to LDC's with little modification. Associated with these techniques is a dualistic economic structure with a high wage, low employment sector co-existing with widespread unemployment (and perhaps an 'informal' sector where much lower wages and higher labour intensity prevail). The highly paid 'formal' sector workers demand sophisticated 'modern' consumer goods. These themselves necessarily also require high capital intensity (13) and reinforce the unequal distribution of income, high unemployment and dualism. Backward and forward linkage investments too are of a capital-intensive nature, it is argued. Growth then depends increasingly on demand from a small section of the community. The government finds it difficult to break the vicious circle, and indeed may not wish to do so if it represents the urban élite. Kenyan manufacturing has been cited as an example of this process (14) but a similar pattern has been depicted with respect to primary commodities in the oil industry of Venezuela for example (Mikesell, 1971). Thus dependence on western technology increases and the technology is a force making for 'underdevelopment'.

Two questions arise immediately about the 'inappropriate technology' position. First, how limited really is the range of technical choice available? Work by Howard Pack (1974) over a wide range of countries and products found that frequently there did exist efficient (i.e. 'undominated') labour-intensive techniques. This is especially likely since many products are produced in a series of sub-processes, some of which may be susceptible to the substitution of labour for capital. This view is generally supported by the evidence in a recent survey by White (1978), though the studies he cites, like Pack's, are confined to manufacturing. For mining, no similar survey exists, apparently, though a few 'appropriate technology' studies have been made which show a range of choice (e.g. Killick 1973 and Thoburn 1977a). Stewart's recent work (1978) while not suggesting that the available technique range is always small, does suggest strongly the existence of 'packages'. As mentioned above, the choice of a particular product may determine the use of a particular main process and of particular linkage processes. Again, there is little evidence yet available for minerals, but on a priori grounds one may doubt the contention that the techniques in linkage industries are similarly limited. There is no reason in general why industries which are 'complementary' (e.g. motor car assembly and the manufacture of paint to be used for motor cars) should be 'similar' in their production characteristics (Richardson 1972). Thus, it will be shown that in Malaysia about a quarter of the capital equipment for the capital-intensive tin mining sector is made locally by a highly labour-intensive engineering industry.

Second, if efficient labour-intensive techniques do not exist, why is technology not then modified by firms operating in LDC's; and further, what are the incentives to use capital-intensive techniques when there is, in fact, a choice? With regard to the latter case, an obvious explanation is that factor prices are 'distorted'. As already argued, this may be the consequence of previously adopted 'inappropriate' technology and one therefore needs to enquire further about the motivation of firms. For multinational companies, profit maximisation in a particular LDC is part only of a global profit objective. Horizontally integrated firms may wish to use a standard technology everywhere, in the belief that the costs of adaptation outweigh those of conforming to local factor supplies. Also, in order to maximise global profits, even with unaltered techniques, vertically integrated international firms may fail to locate in LDC's parts of a process (i.e. linked industries) which might be suitable for local factor proportions. Intermediate goods may be purchased from a home country subsidiary to maintain its capacity utilisation, and mineral processing may be located in the home country to mix sources of supply.

Even concentrating on profitability within an LDC, a firm may prefer capital intensive technology. One reason is that large (and especially migrant) labour forces (as in some mineral industries in Africa) may be hard to control. Labour turnover may be high and this may cause labour costs to be understated by a simple calculation based on wage rates. Firms may feel too that a small, closely supervised labour force is less susceptible to unionisation, though the reverse could apply in the sense that a small group of skilled and unionised workers in a capital-intensive activity could bring production to a halt more effectively than a larger, less integrated workforce. The sheer unfamiliarity with working with a large, local labour force is likely to influence foreign firms, though local firms also may try to use capital intensity as a means of labour force control. Control is increased too by the existence of unemployment elsewhere which may result from the general use of inappropriate technology. Fear of rising labour costs also influences the choice, and Page (1976, p.239) quotes the chairman of Rio Tinto Zinc to the effect that 'it is in our view unjustified to start a mining complex on a labour-intensive basis except under the most unusual circumstances. The problem with labour-intensive operations is they tend, as the economy of the country grows, to become uncompetitive'.

In the past especially, a high capital intensity (in the sense used so far of a high investment cost/workforce ratio) was sometimes used as a weapon by foreign firms to drive out local competition, since high capital intensity often was associated with economies of scale and with high initial capital requirements, which local firms found difficult to finance. Even today, when local LDC firms have better access to capital, high capital intensity combined with very sophisticated technology can still be used as a barrier to entry.

In minerals, barriers to entry appear to vary greatly with the degree of vertical integration. Where firms seek to expand horizontally, for example to exploit the rent obtainable from their technical expertise or management skills, (15) diversification overseas is unlikely to preclude locally owned production in the host-countries. Local mining companies will tend to sell to independent buyers and local entry would only be prevented if the necessary mining technology is too complex or capital-intensive. McKern (1976, pp. 12-14) notes that indigenous companies in the booming Australian minerals industry are well represented in coal, iron ore, lead-zinc, beach sands and tin - where vertical integration is slight. In contrast, in industries which are highly concentrated and oligopolistic, manufacturing firms have a strong incentive to integrate backwards to secure their sources of supply; and in bauxite, petroleum, nickel and to some extent copper, only large multinational companies tend to be represented (McKern, 1976, p.13) (16).

Most discussions of appropriate technology are concerned with employment creation, income distribution and the alleviation of poverty, in the belief that economic growth accompanied by continuing (or rising) unemployment and inequality cannot be considered to be 'development'. Important too is that notion that systems using inappropriate technology tend to reproduce themselves (i.e. the process is cumulative) and also to be excessively depend on western countries. This leads in turn to a poor bargaining position and loss of the gains from foreign investment. In these circumstances local control of production and technology is likely to be an important weapon in the securing of the gains from export expansion (17).

Returning to the trade and development literature, and to the staple theory in particular, it is worth noting finally that there is in it a much stronger thread of technological determinism than in the later literature on technology. Technological differences between products are thought by staple theorists to condition not only some of the effects of export expansion but also the type of ownership (which itself further influences the development effects)(18). Some products therefore are more suitable for foreign firms than are others. Similarly, the technology of linkage industries will condition whether they are located in an LDC or not; and if so, by whom they are produced (19). What is being argued in the staple theory is not that input-output coefficients are fixed for each industry, but that there are basic differences between products in the elasticity of substitution of labour for capital and in economies of scale. The 'appropriate technology' literature in turn highlights the proposition that foreign firms not only have a strong incentive to use capital-intensive methods, but that their market position can often be improved (or indeed established) by the development of capital-intensive methods where these previously did not exist (20). The question then is whether some products resist attempts to develop more capital-intensive methods to a greater extent than others - and are therefore less likely to come under foreign control or to generate a lopsided pattern of development.

Certainly most minerals have proved highly suitable to increasing capital intensity (21). In some cases, though, a small scale local sector has co-existed with a large scale foreign one and there has sometimes (as in tin) been local entry into the foreign technique. Nevertheless the enormous initial costs of many mineral operations, based on the high capital intensity required to exploit low grade ores with large throughputs, has given multinational mining corporations an increasingly strong position *vis-à-vis* potential local rivals, and even LDC national mining organisations (whose role has also increased) frequently use MNC's as contractors.

The appropriate technology literature thus yields some useful insights into mineral exports and growth. One important lesson from the staple model could be applied though to work on appropriate technology. This is that an important role of new activities should be to alter as well as adapt to local factor endowments. A demand for moderately skilled labour, for example, will allow the quality of the local workforce to improve, and perhaps provide workers with skills and savings to set up in business on their own later.

## 2.3 MEASURING THE BENEFITS OF FOREIGN INVESTMENT AND ITS ALTERNATIVES

In the light of the preceding discussion, and with given export prices, potential further gains from exports and investment can be grouped conveniently under three headings:

1. The raising of *retained value* (RV). This is the share of a given export income which accrues to local income recipients. It involves backward integration into the production of intermediate products or capital goods, increasing taxation or otherwise reducing the share of foreign profits and other foreign factor incomes.

2. The raising of the country's share in the export good's *final consumer value* (FCV). This involves forward integration into processing or the manufacture of final products, or increasing local participation in marketing and distribution.

3. The stimulation of wider development effects, especially employment creation, labour training and favourable

income distributional effects.

These approaches are also relevant in assessing the benefits which have already accrued to the host economy through market forces and the past effects of policy. Beyond these approaches it will be extremely useful if a real income measure, or measures, can be developed with which to evaluate alternatives, both at the project and sectoral level. Some recent work on manufacturing investment will prove relevant in this regard, particularly that of Lall and Streeten (1977) and Bos, Saunders and Secchi (1974), as well as the existing studies of mining investment.

In empirical work on trade and development it has been especially important to determine the distribution of gains from exporting, first between local and foreign income recipients, and then (within the latter category) between workers, profit receivers and the government. The local/foreign split is especially relevant to the notion of enclaves, where real income is thought to flow out of the host economy with little domestic effect. In consequence, the calculation of retained value is a natural starting point. It has been used for instance in the collection of studies of petroleum and mineral investment by Mikesell (1971) and his associates and in Mikesell's more recent (1975) study of copper investment, and it is now a common tool in commodity studies generally (22). RV gives a measure of the direct effect on the balance of payments and allows income flows to be identified whose further progress through the domestic economy can be traced. Thus linkage purchases can be traced and the local content of workers' expenditure calculated, given sufficient further data on consumption patterns.

RV as a balance of payments measure is useful mainly if one thinks that the LDC concerned has a'payment-gap' kind of constraint on its foreign exchange receipts, such that balance of equilibrium cannot be secured by domestic policies. Even in that case, though, indirect effects on the balance of payments via imports by local income recipients and the government would have to be considered too. Also, if the export activity is large (i.e. not 'marginal') in relation to the domestic economy the balance of payments effects of induced changes in the exchange rate and the domestic price level would be relevant and would affect other sectors' external payments position. More important is the potential use of RV as a real income measure, in the sense of its representing free foreign exchange (i.e. command over foreign resources) in the hands of the government and local people. Here too, though, modifications must be made to the RV measure if the export activity is non-marginal (see later in this section).

Leaving aside for the moment the possibility that the export activity is not 'marginal' to the economy, the identification of RV with a real income gain from exports needs some qualification. It holds true only if the domestic resources employed in the export sector have no alternative use. This would imply that all labour used was previously unemployed and could be allocated a zero shadow wage (which is unusual (23)); and further, that any local investment was financed entirely out of newly created savings (i.e. in response to the investment opportunity - Hirschman, 1958, p.37) or that the economy was so short of investment outlets that the savings used would otherwise be 'wasted' (Streeten, 1961, p.64).

If foreign investment is involved and if local investible funds have alternative productive uses, one must ask whether the foreign capital inflow adversely affects domestic savings. This issue was raised in the early years of the trade and development debate (e.g. Meier, 1953, 1968) and further work has been done on it in connection with private foreign investment in manufacturing. Reuber (1973, pp.36-37), following an earlier study by Weisskopf (cited by Reuber, p.36), regressed a proxy for locally financed domestic investment (i.e. Total Domestic Investment - Foreign Investment) on foreign private capital inflows using time series data for fifteen countries and cross sectional data for twenty-five. The finding was that in the majority of cases the relation was positive and statistically significant, reflecting the probability that at least domestic investment had not been discouraged by the foreign capital inflow. In the analysis of foreign direct investment, much emphasis must be placed on the hypothetical situation where no foreign capital inflow occurs. Besides the effect on the level of domestic investible funds, researchers have asked whether, in the absence of the foreign firms, local firms would have produced the output (24). It would be possible indeed to hypothesise a situation where the foreign investment prevents the setting up of an identical domestic firm, and where the domestic investible resources which would have been used are instead remitted abroad. Here there is a loss to the economy equal to the increase in the proportion of current output value spent abroad by the foreign firm (e.g. on profit remittances).

Whether, with a given level of domestic investible funds, a new private foreign investment replaces (or prevents) local investment often in practice depends on technology and on the degree of vertical integration, as suggested in the previous section. If local entry is impossible, then the foreign investor's output is purely additional. If local entry is possible, but has been prevented (or pre-empted) by the foreign firm, than an alternative situation must be hypothesised where the net gain from the foreign investment is the difference between these situations. Lall and Streeten (1977) have carried out analyses of this kind in some detail, covering manufacturing investment in six developing countries. Since much of this investment is for the domestic market, they are strongly interested in moving from a direct (i.e. an RV) measure of the balance of payments effect to one which is 'total', and which is related to a measure of social income, but their methodology is relevant to mineral export investment too. Three alternative situations are considered. First, it is assumed that the entire output of the investment would be imported (i.e. if not produced by the foreign firm). This allows the output to be expressed in terms of foreign exchange, so giving an import substitution project an equivalent value to an export-orientated one (25). This allows 'total' balance of payments effects and the social income effect to be evaluated simultaneously. Alternatively, total local replacement might be possible, in which case (taking other benefits and costs as given) the costs of servicing the foreign investment can be compared to the opportunity cost of local capital (26). This isolates the effects of the 'foreignness' of the investment, though only in terms of financial resources, not improved management, technology, etc. (Lall and Streeten, 1977, p.164). Third, an impressionistic estimate is made of the degree of possible local financial replacement likely in practice (i.e. if the investment opportunity had not been pre-empted by the foreign firm). The resulting estimate of social income for a given time period (i.e. with the capital costs of local investment treated as an annual charge) can be expressed as a percentage of annual sales value.

To some extent the Lall-Streeten approach is little more than a simplification to a single time period of the Little-Mirlees method of converting costs and benefits into foreign exchange, in order in this case to move from the direct balance of payments effect to the total balance of payments and social income effects (27). Given sufficient data it could be extended to cover net benefits over time, so that a net present value for a foreign (or domestic) investment could be calculated (28). In consequence the methodology used in chapter six of this report to appraise the benefits of new tin mining investment uses a full cost benefit analysis, (29) taking into account the alternative of domestic entry.

It is worth pointing out that Lall and Streeten are not alone in converting RV into a measure of net social gain. Pearson and Cownie's (1974) series of case studies of commodity exports from African countries use a net gain coefficient where net social gain (similar to that of Lall and Streeten) is normalised by dividing by the value of exports at the shadow exchange rate (30).

Lall and Streeten's contribution is principally to spell out the alternative situation more carefully, which is as relevant for export-orientated as for domestically orientated investment. In practice too (as in Pearson and Cownie) the net social gain from backward linkage investment is not considered because of lack of data. In other words, intermediate products are simply valued at border prices if traded, and using a Little-Mirrlees 'standard conversion factor' if non-traded, and the possibility that (if non-traded) they might use non-scarce labour or newly created investible funds is not investigated. This practice is also followed in chapter six, but the linkages are described in chapter five, which also gives estimates of the net gain from existing production. The Lall-Streeten method is of course a static measure of the gains from exports and foreign investment, though the increased investment funds thus made available can be cumulated to raise the growth rate. It may be noted in passing that the local expenditure figures for export-orientated manufacturing are often substantially below the RV figures for many mineral exports. Thus a study by Reuber (1973) which covers a wide sample of US, European and Japanese manufacturing investment overseas, estimated that overall cash expenditure in the host economy for export-orientated projects was only 34 per cent of sales revenue (31). Gillis (1978, p.10) suggests RV figures of under 50 per cent for most foreign mining projects (compared to 50-80 per cent for locally owned mining), but the mineral studies cited in Thoburn (1977b, chapter ten) contain evidence of somewhat higher figures.

While most studies deal with projects marginal to the economy, the aims of Bos, Sanders and Secchi (1974) are more ambitious. They claim they can assess the effects of private foreign investment on income gains and on the rate of growth even of activities which are not marginal. In my view their attempt, though interesting, is largely unsuccessful, mainly because they have to exclude by assumption so many important aspects of the problem in order to set up a formal model. Nevertheless it has had some influence on commodity studies, (32) and an attempt must be made at least to justify its dismissal.

Bos et al. are interested in the alternative situation where no foreign investment has occurred, in relation to the 'total' effects of the investment. They rightly stress the need to consider the balance of payments implications of domestic income increases caused by capital inflows. There is not space here to set out their model but basically they treat the sum of the balance of payments current account deficit (equivalent, with given foreign exchange reserves, to capital inflow), private savings and public savings created by the investment, as the addition to domestic investible funds. This addition can then be calculated in incremental form, or fed into a simple growth model, in order to compare the new situation to that which would have occurred without the investment. Certainly, 'total' effects are considered in the income changes react back on the other determinants of investment funds, but the level of domestic prices, the terms of trade, and the exchange rate are held constant (Bos et al., 1974, pp.90-96). Yet it is precisely in these areas that a non-marginal foreign investment would differ in its effects from a marginal project. In particular, large capital inflows may well cause pressure on the exchange rate to appreciate, with implications for the terms of trade. Further, the assumption that the current account deficit can be identified with capital inflow leads the model to suggest that a shift towards foreign investment making a greater contribution to import replacement (or export expansion - Bos et al., pp.71 and 223-26) is harmful. Thus a project which (abstracting from the initial capital inflow) tends to improve the current account balance of payments, actually tends to lower social income in their system (33).

The exchange rate changes resulting from capital inflows mentioned in the preceding paragraph are also likely to occur as a result of export expansion itself. One attempt to estimate the exchange rate changes resulting from the past expansion of mineral exports (and much more interesting than the Bos model) has been made by Jenkins and Gillis (in Gillis, 1978, chapter three), who also assess (for 1973-74) the effects of the exchange rate change on welfare in Bolivia (34). In principle the sudden and hypothetical loss of the country's mineral exports would cause the exchange rate to depreciate. This would tend to encourage non-mineral exports (unless their supply elasticity were zero) and to reduce import volume. The welfare losses from not having a minerals sector are then measured as the consequent saving in resource cost of producing minerals less the resource cost of expanding non-mineral exports and the welfare loss of reduced imports. The new equilibrium exchange rate is estimated on the basis of assumed values of the two export supply elasticities and the domestic price elasticity of demand for imports. The world (dollar) price of Bolivian exports is assumed to be fixed and similarly Bolivian import purchases leave unaltered the world price of her imports. The attempt shows that Bolivian mineral exports generate a gain in producer and consumer surplus equal to about a third of the export revenue earned by the mining sector.

The Gillis and Jenkins method does not deal with non-traded goods explicitly. They mention (pp.84-85) that devaluation causes import prices to rise but since the international terms of trade are fixed by assumption, such a rise cannot take place relative to exports, only in terms of tradeables (imports and exports) against non-traded goods. Their measure of the welfare loss from reduced imports is thus one of the change in total import value (assumed equal to the change in total export value) and the loss of consumer surplus from reduced imports. No consequent changes in non-traded goods production are considered. It is also worth noting that they see the gains resulting from exchange rate changes due to mineral exporting mainly in terms of the reallocation of existing resources. Work of a related kind, which also assumes fixed international terms of trade, has been carried out with regard to Australian mineral expansion. A partial equilibrium analysis by Gregory (1976) has shown that the exchange rate appreciation resulting from mineral export expansion has harmed non-mineral (mainly agricultural) exports and has had an equivalent effect to a large tariff reduction on import competing industries (35). Snape (1977) has provided a general equilibrium restatement of Gregory's analysis of the need for adjustment in the non-mineral sectors. This also takes non-traded goods into account and shows that gains from mineral exporting result from an outward shift of Australia's production possibility frontier. Capital movements also can be taken into account in the analysis, though to some extent the investment inflow into Australian minerals has been counter-balanced by an outflow in the form of imports of capital goods and intermediate products for mining.

Returning to export investment which does not have any significant impact on domestic prices or the exchange rate, some further points of caution are necessary about using increased RV as a measure of additional gain. First, where taxation of foreign profits on existing production is a method chosen, the resulting increase may be at the expense of future investment. This difficulty is best discussed in the next section, where taxation is considered in relation to resource rents and other rents. With regard to backward integration, the possibility of some backward linkages being 'bad' ones is now well established. Clearly not all potential backward linkages could develop independently of local purchases by means of direct export. This is especially true of specialised mining equipment where local customer contact is important. Nevertheless the net social benefits must be examined in each case before proceeding, if only in a rough and ready fashion, and sometimes they may prove negative (e.g. because the activity is subject to great economies of scale or is inappropriately capital-intensive). On the other hand, appropriate local linkage development <u>may</u> have been precluded by the purchasing policy of foreign-owned firms. This is considered below in connection with processing.

More generally it is always worth asking to what extent a particular RV figure depends on the nationality of ownership, and to what extent on the productive technique being used. Sometimes where local mine ownership is possible, choosing to promote local rather than foreign investment is also effectively choosing one technique rather than another. While there are few cases like Killick's (1973) classic study of diamond mining, where the foreign company used highly capital-intensive methods, and local miners worked almost by hand, often (as section 2 showed) technique and ownership are closely related. This is true in tin. Hence a government has the choice between raising RV by adopting a more stringent policy towards foreign production and investment, or promoting local sector growth on the assumption that its RV is higher.

An aim often expressed in international forums, and especially as part of the New International Economic Order demands promoted by the United Nations, is that LDC's increasingly should process further their own primary exports and move into the production of final products. This kind of export-orientated Resource Based Industrialisation thus is a call for increasing the local share in Final Consumer Value by forward linkage investment. Partly this is advocated in the belief that it is a convenient way for LDC's to 'bite-off' additional chunks of value added, but also that it improves their bargaining position in that substantial economic rents accrue at the processing stage. Clearly in some cases bargaining power is improved by forward integration by local firms or LDC state corporations (36). Hughes and Singh (1978) suggest that tin smelting has been a method where rents have been shifted to Malaysia away from consuming countries. Widyono (1977) has chronicled how from 1953 to 1961 the Williams Harvey smelter in England (a smelter then owned by Patiño) imposed a levy on Bolivian tin concentrates as compulsory compensation for the nationalisation of Patiño's tin mines in Bolivia. Generally the buyers of concentrated ores are fewer than the buyers of fabricated metal so sellers of the latter face less monopsonistic power. To go beyond these general points about bargaining, (37) it is necessary to consider the degree of vertical integration of the foreign mining companies in the metal in question. Factors making for vertical integration have been discussed in the previous section, as have the conditions for local entry into mining. Where there are local mining companies or where the state engages in mining, it is possible that the international market even for fabricated metal may be controlled by the mining companies. The structure of integration in the tin industry is discussed in the next chapter but it is worth mentioning here that there is little vertical integration in tin companies and that most smelting is done in the LDC producing countries themselves though marketing is often in the hands of foreign traders. In other industries, such as aluminium, it is necessary for LDC's to integrate forward beyond the metal stage in order to escape the market power of vertically integrated multinationals.

Apart from a potentially improved bargaining position from increasing the domestic FCV share, there seem considerable grounds for doubting the value of RBI as an industrialisation strategy. In a recent World Bank study, Wall (1979) argues that RBI has failed to occur as a result of the appropriate operation of market forces rather than as a result of any artificial barriers (which might be removed). He disputes that the structure of protection in industrial countries is biased against processed goods, (38) and that freight rates artificially discriminate against transporting processed products. Roemer's very useful survey (1979) of RBI is also sceptical. Roemer notes that most resource-based industries have capital/labour ratios higher than the industrial average in developed countries. Weight loss in processing can sometimes counteract this especially in ores where the metal content is low, and other inputs such as energy may dominate costs so as to render irrelevant the capital/labour ratio as an indication of 'appropriateness' for LDC conditions. Note though that metal using industries typically have rather low K/L ratios, so LDC's stand to gain if they can integrate beyond the metal processing stage (Roemer, 1979, p.171). It is also worth noting that this is a point in favour of local capital goods production in LDC's which of course is heavily reliant on metal using industries.

The bargaining element present in raising RV in existing foreign owned production (or in the prospective operating stage of new foreign investment) is relevant too in attempts to increase the local share in FCV the more integrated vertically are the mining companies involved. In vertically integrated industries the interests of the mining companies and the host-country can well diverge (39). Thus the companies may wish to continue processing in their home countries to keep existing capacity utilised, to mix minerals from various sources or to reduce the risk of expropriation. Where independent overseas processors are involved, the LDC position is stronger, unless the processors themselves are owned by oligopolistic metal users. The structure of the tin industry in these regards is discussed in the next chapter.

In general, similar methods of appraising the social desirability of forward linkage investment can be used to those for backward linkage. Their appropriateness to local conditions can be indicated by their capital-intensity, taking into account the role of any 'dominant' input other than capital and labour. Further, though it is not done in this study, forward linkage investment can be appraised by cost-benefit analysis provided realistic estimates of expected prices can be made.

The third type of method for raising the contribution of mineral exporting to development - that of stimulating wider development effects - has less to do with policy towards foreign investment per se and more to do with foreign investment in relation to local production. This is because the scope for using negotiation to increase local employment and the wage structure is limited, (40) and effectively it may be necessary to use local firms (or state firms) if the productive technique is to be changed (41). Although it is useful to indicate employment creation and income distributional effects through the wage structure directly, both can in principle be included in net present value measures. This is done in chapter six.

## 2.4 MINERAL DEVELOPMENT POLICY

It has been shown that an analysis of the development effects of mineral exports is in many respects similar to that of primary commodities more generally, and work on the benefits of private foreign investment in manufacturing in LDC's has also been examined for its relevance to mineral development issues.

Minerals are distinguished from other products, however, in that they are exhaustible, at least in principle. It is therefore particularly important that mineral policy be geared to the eventual construction of economic activity which is self-sustaining after mining has ceased. In the words of a recent 'radical' study of African mining - 'When the companies have exhausted Africa's major mineral deposits, will the continent have anything more to show than huge holes in the ground?' (Lanning and Mueller, 1979, p.24). Of course, there are no good reasons for assuming an infinite economic life for <u>any</u> form of productive activity. Changing tastes may make individual manufactured products redundant and changing relative cost patterns may lead to a complete relocation of the activity away from a particular country. Agricultural products too may lose their markets if effective, lower cost substitutes are developed. Conversely, exploration leads to the discovery of new mineral deposits, and advances in extraction technology can turn previously unexploitable mineral resources into reserves of economic value.

Nevertheless at any given time there will be only finite known reserves, and the appropriate rate of depletion is considered first in this section. Tax policy, the second topic, is one important influence on depletion decisions. It is also often given special consideration in the belief that, especially for foreign investment, market forms of development such as linkages and employment creation are slight in mineral projects, so that taxes become the main source of increased retained value. The scope for taxation is sometimes greater in mining ventures if rich deposits are mined simultaneously with poorer ones such that the former generate returns in the form of resource rent over and above the supply price of investment. The monopolistic and oligopolistic structure of mining industries gives rise in addition to monopoly rents which can be taxed also, and the incidence of taxes varies with market structure. Tax policy can, of course, be combined with provisions to strengthen other development effects so as to reduce further the enclave character of mining. Also important, and considered last, are alternatives to tax policy such as equity participation, management contracts, and production sharing.

2.4.1 The rate of depletion

The fundamental notion of optional resource depletion is that mineral deposits represent an asset to their owners, whose value in equilibrium over time should grow at the market rate of interest (42). Thus given an initial net price, $p_o$ (market price less extraction cost), current net price, $p = p_o e^{rt}$. This upward tilt to the price path is then associated with high outputs in the early periods, which subsequently declines and the resource is exhausted at exactly the point where it has priced itself out of the market (Solow, 1974, p.3). More formally, under perfectly competitive conditions, price would equal marginal extraction costs + user cost, where user cost represents the fall in the value of the mineral asset brought about by the extraction of an extra unit - a kind of depreciation allowance for the mineral deposit in other words. User cost must then grow at the market rate of interest, and the time path of market price then depends additionally on the time path of extraction costs. Under monopoly, it is marginal revenue which is equated to marginal user cost + marginal extraction cost (Peterson and Fisher, 1977, pp.685 and 693-94). Much theoretical debate has taken place on whether monopoly leads to a lower rate of extraction. Hotelling's (1931) paper, using a linear demand function, showed it did, and he attributed this to the general tendency of monopoly to restrict output. Later work has disputed this, using constant elasticity demand functions (e.g. Stiglitz, 1976) but even Stiglitz concedes that other assumptions (e.g. rising demand elasticity over time as substitutes for the mineral ore developed) will lead to longer mine life under monopoly. Even under perfect competition the depletion rate will not be optimal if the market rate of interest used to determine the time path of net price differs from the social rate of time preference. Generally the latter will be the lower since market rates take account of risk and relate to after-tax returns (Solow, 1974, p.8).

In principle the socially optimal rate of depletion of a given mineral deposit is that which would maximise its social net present value. Social NPV would be arrived at by using the social time preference rate of discount and, if necessary, by substituting other shadow prices for market prices (though this is not normally mentioned in the optional depletion literature).

Many practical problems face a producing country deciding on a depletion rate. The discovery of new resources of its own might lead it to need a faster depletion rate than originally seemed desirable, and the discovery of new reserves elsewhere or the development of substitutes would affect price expectations. Mining technology often requires heavy fixed capital equipment which would normally be run to capacity. This generates a need for relatively steady rate of production, as opposed to the gradually declining rate generally thought to be optimal (43), (44).

Frequently mineral deposits vary in richness and in accessibility, so that extraction costs differ. Differences of this kind are especially associated with underground versus open cast or alluvial deposits, and (as in the tin industry) a particular kind of deposit may occur predominantly in one country and a different kind in others. In theory, such variations in extraction costs should lead to the lower cost deposits being exploited first, and price would rise to cover exploitation of higher cost deposits only once the former had been depleted. In fact, high cost and low cost mining frequently co-exist, and they certainly do so in the tin industry. If the countries owning low cost deposits choose not to exploit at a rate great enough to force price below that necessary to keep high cost producers in operation, the low cost producers earn substantial resource (or mineral) rents, analagous to the Ricardian rent which arises from farming intramarginal tracts of land. Such rents are best considered with tax policy.

In view of the uncertainties about future prices and reserves, this study does not attempt to estimate optimal depletion rates, and it is worth noting that there is little evidence that the secularly rising net prices predicted by depletion theory have occurred in practice over the last hundred years (Peterson and Fisher, 1977, p.708). Chapter six on the gains from investment in tin does consider in an ad hoc fashion some depletion decisions, however, and it is worth noting that the maximum NPV of a deposit, with respect to mine life, is sensitive both to price expectations and to the discount rate. In general a rise in price tends to shorten project life (see Appendix). A rise in the discount rate also tends to raise the depletion rate of a given deposit but by its discouraging effect on investment it could result in a fall in prospective output for a country as a whole.

Finally, as Hughes (1975) has pointed out, a developing country may also wish to base depletion decisions on its capacity to absorb the flow of income and taxes which the mining generates. The responsiveness to linkage investment opportunities offered by mining projects tends to rise over time (45) and the government's ability to use tax revenue for development purposes may also do so. The alternative is for returns to be invested overseas, the policy followed by some of the middle east oil producers; but this carries risks of expropriation and unforeseen capital loss.

### 2.4.2 Taxation policy and resource rent

The role of taxation as an influence on retained value has already been mentioned in general for foreign investment in the previous section. It is sometimes argued that mineral projects are typically such enclaves (i.e. generating so few linkages and so little employment) that in the absence of tax payments RV would be extremely low (46). Taxes are important, too even on domestic firms if it is believed that the mining companies concerned are so specialised that they would automatically reinvest their profits in other mining (perhaps elsewhere in the world) rather than in more diversified activities which would outlive the exhaustion of the resource. This is more an argument, however, which relates to rich mineral producers such as Canada and Australia, where mining is to some extent undertaken by home based transnationals who are specialised in mining. Where there exist local miners in LDC's (such as the Chinese in Malaysia and Thailand) there is much less reason to believe that profits would be reinvested in mining (47). State mining companies also are generally taxed on the grounds that the government is likely to make better use of their surplus than they would themselves, and past scandals concerning the investment policies of the Indonesian state oil company, Pertamina, have reinforced this view in South East Asia (Gillis, 1978, p.10).

An additional and distinctive dimension of tax policy in minerals concerns the widespread existence of economic rent, in the classical sense of a return to the factors of production employed over and above their supply price. This is usually applied to the return on capital though it can 'leak' to other factors and to the suppliers of intermediate products. Such rent can in principle be taxed in its entirety by government without reducing the incentive to further investment. Resource rents of this kind arise from the existence of intramarginal producers in the Ricardian sense, who can extract at costs below those of the marginal producers, when market price is high enough to keep the latter in business. User cost, though referred to sometimes as a 'scarcity rent' (e.g. by Solow, 1974), is normally taken to be part of supply price and is not thought to be taxable without affecting factor supply. Since it represents a depreciation allowance for the diminishing value of the deposit over time (and tends to grow over time) it would seem inequitable to tax it if the mining company had purchased the deposit, and to do so would in any case at least damage the incentive to undertake expenditure on any further exploration (Steele, 1967, pp.238-39). It accrues even on marginal deposits since the marginal producer has the option of waiting for net price to rise (at a rate equal to the rate of interest) and thus will only produce once price covers marginal user cost as well as marginal extraction cost (48).

As well as resource rents, monopoly rents arise in practice in the highly integrated structure of world mining, and tax policy is also aimed at securing a share in these (49). Market structure in turn is an important determinant of the incidence of taxes (i.e. who effectively pays them). Dominance of the market by a producing country or group of countries is one way that export taxes or royalties can be passed on to the consumers, though of course the incidence depends on the price elasticity of the product's market and the price elasticities of supply. Monopolistic producers will tend to absorb more of an output tax themselves than will firms under perfect competition (50).

In principle, taxation to extract resource rents sounds straightforward. In operation it is complicated by the existence of monopoly rents and quasi-rents to specialist factors. Taxing quasi-rents may well lead to a shortage of specialised management, for example, and there is great difficulty in determining in any case what the supply price of investment really is. Mining projects are usually rather riskier than manufacturing ones, particularly when one takes into account the costs of potentially unsuccessful exploration. Owing to the long term nature of much mining investment, and its high capital intensity, very high taxation may be possible in the medium term (in the sense that firms can still generate an after-tax operating profit) but this may well be at the expense of long run investment.

The structure of taxation is important too in apportioning risk. Taxes on output, which are levied whether or not an operation is profitable, place risk on the company, whereas profits taxes do not. One interesting proposal to reduce risk and to secure a large share of resource rents is to levy a resource rent tax (RRT) on profits over and above a specified target rate of return (Garnaut and Clunies Ross, 1975). Profits are allowed to cumulate at this target rate and once an excess is built up it is taxed. The tax rate is not 100 per cent however since this would destroy the company's incentive to efficiency and might result in a net fall in the tax take.

Taxes also may affect depletion rates. In particular an output tax has a similar effect to a fall in expected price and thus shortens mine life, while an income tax in principle should not affect this (51).

### 2.4.3 Alternatives and supplements to tax policy (52)

Traditionally, mining companies in LDC's have been subject to output or export taxes or royalties and to income tax. Export taxes often date back to colonial times, being easy to collect and difficult to avoid, and taxes on profits are usually of post Second World War origin.

In some countries foreign mining companies are subject to uniform tax regimes, at least for a particular mineral, and these are often variants of the tax system for industrial enterprises. In other cases agreements may be negotiated individually. Both general mining legislation and particular mining agreements can include economic development provisions of the sort implicit in the earlier sections. Local purchase of supplies can be encouraged or stipulated, as can further processing, the training of nationals in management posts, the provision of infrastructure, and so on.

Linkage investment, including processing if it does not occur naturally in response to market forces should be examined carefully to see whether its promotion is appropriate for local conditions. Companies may be reluctant to buy locally what they can obtain from overseas affiliates or to process locally if this is felt to reduce their market pwoer. Hughes and Singh (1978) for instance have shown that considerable economic rent accrues in many minerals mainly at the processing stage, and has not been appropriated by LDC's, who continue to export the unprocessed product.

Even with uniform tax and economic development provisions for all new entrants, the amount of foreign investment will depend on the relative bargaining positions of the host-country and particular foreign companies. Thus stringent provisions may result in the entry only of companies in a fairly weak position - those, for example, seeking a secure source of resource supply for some downstream activity and lacking it at present. Similarly, the market power of established companies might be such that provisions to secure further processing, if this were economically desirable, might be difficult to introduce.

Smith and Wells (1975, chapter two) distinguish between 'traditional' and 'modern' concessions. 'Modern' concessions go beyond the combination of tax and economic development provisions outlined above typically to give the host government a greater direct influence in the mining operation, and they mostly date from the 1960's and 1970's. Government equity participation is often sought on political grounds as a 'moderate' alternative to nationalisation. On economic grounds state participation can be attacked because a properly functioning tax system should be able to secure for the government all the benefits from resource rents and from monopoly rents, leaving the foreign investor simply with the supply-price of his investment. The difficulties of designing a resource rent tax have already been discussed, however, and the uncertainty about the size of the feasible long run tax take may persuade governments to take an equity stake. In addition, a government short of investment outlets might like to earn 'normal profit' from a mining investment, like any other, and yet wish to share the venture with a foreign investor of established expertise. Foreign investors, too, might welcome the government as a partner to reduce the risk of nationalisation, and the reduced risk *per se* might lead the company not to deplete at an inappropriately fast rate to ensure a short pay back period. The state as a shareholder, and perhaps as a management partner too, might also expect to influence such decisions as local purchasing policy, though it could also do this through legislation.

Other forms of 'modern' concession include management contracts where an industry has been nationalised but where a foreign company is invited to operate the mines. Production sharing arrangements involve the host government retaining some share of the output of the project, and have been common in the oil industry. Other variations include service and work contracts where the foreign investor essentially works as a contractor, receiving his costs and a share of the profits.

NOTES

(1) For a more general account of trade and development, see Thoburn (1977b, chapter three).
(2) See in particular Singer (1950) and (e.g.) Myrdal (1957) and Prebisch (1963).
(3) At present (1977 statistics) primary commodities generate 82 per cent of LDC export earnings, though this drops to 57 per cent if petroleum is excluded (World Bank, 1979).
(4) Note that in fact LDC's produce only about a third of world (non-oil) primary commodity exports, but it is with LDC primary commodity exporters that the literature is concerned. The LDC share rises to nearly 60 per cent if petroleum exports are included (World Bank, 1979).
(5) If resources are transferred to alternative uses, it makes empirical tests of deteriorating terms of trade unreliable. A tendency towards deterioration could have occurred, but has been forestalled by resource reallocation. For a general discussion of the debate on the terms of trade deterioration see Meier (1968, chapter three). A useful more recent account is given in Evans (1979).
(6) In Hirschman's recent (1977) reformulation of linkages he has added to the earlier concepts of backward and forward linkage those of final demand or 'consumption' linkage (used by Watkins 1963 and described above) and fiscal linkage (i.e. tax revenue paid to the government). He notes that fiscal linkage in particular may be at the expense of other linkages.
(7) In general there is an important distinction to be made between 'domestic' and 'world' linkages. See Jones (1976) and Riedel (1976).
(8) The degree of urbanisation associated with export development will also influence local market creation.
(9) See also Thoburn (1977b, chapter eleven) which discusses the relevance of the Baldwin characterisations more generally.
(10) The distinction between Marxist and dependency theories is confusing to the extent that some dependency writers (such as A. G. Frank) work from a Marxist perspective, though many other dependency theorists would not see themselves as Marxists.
(11) See Evans (1979) for a discussion of unequal exchange.
(12) See Watkins (1977) for a statement of the reinterpretation of the staple theory.
(13) The notion of product-determined choice of techniques is set out in Stewart (1972).
(14) See Stewart's chapter in Dimanchi, Routh and Ali-Taha (1976).
(15) See Caves (1971) for a discussion of the motivation behind foreign investment.
(16) McKern's findings only refer to Australia. In other countries state enterprise has been the vehicle for local entry into mining; and such local entry has occurred even in highly oligopolistic and vertically integrated industries such as bauxite.

(17) This point has been made strongly by Hans Singer (1975) in his return visit to the distribution of the gains from investment. Marxists would argue that ownership and control of production and technology, like multinational companies themselves, simply are part of a wider process of international capital accumulation. However, even if one agreed with the Marxist view, it would be well worth studying the proximate effects of securing greater local control over the choice of technique.

(18) In fact, in Hirschman's (1977) generalised linkage version of the staple theory, he suggests technology may even condition the mode of production. Thus sugar technology was closely associated with the institution of slavery.

(19) Hirschman (1977) distinguishes between 'inside' linkages (suitable for local firms) and 'outside' linkages (suitable only for foreigners or the state).

(20) This issue is discussed for the tin industry in chapter four.

(21) Further details are given in chapter three.

(22) See Thoburn (1977b, chapter ten) for a survey of such studies.

(23) This is, as is well known, because labour is rarely totally unemployed in LDC's, especially if drawn from subsistence agriculture, so output would fall if workers move into the export sector unless other workers work harder. (Little and Mirrlees 1974, p.170). Also newly employed workers will increase their consumption and thereby raise consumption in the economy relative to savings - which represents a social cost if there is a shortage of investible funds (Sen 1975, p.100).

(24) This has been of particular interest to researchers looking at projects producing for local markets, where the direct balance of payments effect is often negative and the researcher wishes to know whether the alternative is imports or locally owned domestic production. It can also be applied to export orientated projects.

(25) Obviously for an export orientated project the alternative would be zero exports rather than an equivalent amount of imports (since there is no local demand).

(26) In fact this understates the real cost of using local funds in place of foreign funds. This is because in most LDC's the withdrawal of local investment funds to replace foreign investment will leave some labour unemployed (whose loss will equal the difference between its previous market wage and its shadow wage).

(27) Lall and Streeten's 'total' measure is a real income one. This is converted into a 'total' balance of payments measure by assuming what income accruing to the government and private local firms is entirely saved, while that accruing to local non-scarce labour (which could be spent on imports) is negligible (1977, p.197).

(28) Unless some domestic investment was involved, there would be no point in calculating an internal rate of return since there would be no negative component in the cash flow.

(29) Though the Squire-Van der Tak (1975) approach is used in preference to Little-Mirrlees.

(30) This, in turn, is rather similar to the Domestic Resource Cost measure sometimes advocated as a short cut to a full cost benefit analysis. See Thoburn (1977b, p.53).

(31) The average figure for all types of manufacturing investment was 62 per cent including a figure of 73 per cent for local market orientated projects. Note though the Lall and Streeten findings (1977, pp.172 and 182) that nearly 40 per cent of their sample projects yielded

negative net social income and that these were predominantly in host market orientated projects granted high effective rates of protection. Such a discrepancy between RV and net social gain would not appear in export projects, however, unless they were heavily subsidised.

(32) It has been applied for example to the Indonesian petroleum industry. See Arief (1979).

(33) This is because it lowers investment resources for the rest of the economy, with foreign exchange reserves held constant by assumption (Bos et al. 1974, pp.71 and 333). The context here is similar to Lall and Streeten's assessment of whether a foreign investment project would, in the alternative situation, be replaced by imports or domestic production. Thus in Bos, where alternative domestic production is least likely to be able to replace that of the foreign firm, the latter's contribution to income is least.

(34) Gillis and Jenkins analyse the whole of the Bolivian non-fuel minerals sector, not just tin (which generated 60 per cent of non-fuel mineral export earnings in 1974). Their results are discussed further in chapter five.

(35) In consequence of the adjustment costs for other export industries and for the import competing sector, a government may sometimes wish to prevent the exchange rate appreciating. Such a policy, which would ensure that a mineral discovery would generate a balance of payments surplus has been described by Corden (1978) as 'exchange rate protection'. The policy may also be appropriate if the mineral boom is expected to be temporary.

(36) It also reduces the scope for transfer pricing losses.

(37) The role of market structure in determining the relative bargaining positions of host governments and foreign investors in three major minerals (bauxite, iron ore and copper) is set out in Radetzki (1978).

(38) Yeats (1979, chapter four) in a wide ranging study of trade barriers facing developing countries, concludes that on average effective rates of tariff protection in industrial countries do rise between that on minimally processed commodities to the next stage of processing, but fall thereafter.

(39) See Radetzki (1977) for further discussion.

(40) See also the final part of section 4.

(41) In fact, as the next chapter will show, local firms are usually more likely users of labour-intensive techniques than state corporations.

(42) Work on optimal rates of resource depletion dates back principally to Hotelling (1931) and there has been a great revival of interest since the early 1970's. Hotelling himself draws on Gray (1914) which is still a valuable exposition of the effects of discount rates and prices on depletion and of the meaning of rent. See Solow (1974) and the survey by Peterson and Fisher (1977) for statements of the present state of knowledge.

(43) For discussion of this point see Hotelling (1931) and Lecomber (1979, pp.55-56). This is the justification for the assumption of 'steady production' made in the analysis of depletion in the Appendix.

(44) This is not to say, as will be shown later, that optimal depletion occurs if the deposit is mined over the whole life of the capital equipment. In some circumstances it will be desirable to put in larger amounts of equipment and then scrap it at the end of the life of the ore body (see Appendix).

(45) This is a general point and indicates the tendency of primary commodity and manufacturing enclaves in LDC's to 'break-out' over time. For example, a study of Puerto Rico by Weisskopf and Wolff (1977) showed that the production structure of that economy had 'deepened' considerably over time as linkages had developed. For a general survey of the role of foreign companies in promoting linkages in LDC's with respect to manufacturing, see Lall (1978). Vernon (1973, pp.59-62) also notes a tendency for foreign owned mining 'enclaves' to become more integrated with the host economy over time.

(46) Gillis' influential study of Bolivia takes this view (1978, p.10). My earlier study of tin in Malaysia (Thoburn, 1977b, chapters five and eight) produces some evidence to the contrary.

(47) Historically in Malaya, for example, there has been considerable movement of local Chinese funds between different economic activities (Thoburn, 1977b, pp.67 and 180).

(48) See Gillis (1978) p.98. Gray(1914) denies this, but he is working with a constant price over time.

(49) Hughes (1975) also identifies a category of rent which arises where the mineral has a higher cost substitute. Bosson and Varon (1977, p.180) suggest that price should be set at just below the price of such a substitute and that the gap would represent the intrinsic value of the mineral in the ground.

(50) See Gillis and McClure (1975) for an outline of the principles determining incidence, and for an analysis of tax incidence in the bauxite industry.

(51) For an early statement of this, see Gray (1914, pp.443-45). The effect of alternative taxes on mine life (and cut off grades) are usefully set out in Helliwell (1978).

(52) This subsection draws heavily on the excellent account of mineral agreements in Smith and Wells (1975).

# 3 An Overview of the World Tin Economy

This chapter presents a general view of the industry as a setting for the later work on the changing distribution of benefits from tin exports. Section 1 considers trends in consumption and the pattern of production, and discusses the cartelisation potential of the industry and the International Tin Agreement. Section 2 looks at the available technology in the industry to see whether this predisposes tin towards a particular market structure, though it does not prejudge the issue (dealt with in chapter four) of how technology was developed and for what purpose. Sections 3 and 4 provide introductory descriptions, respectively, of the major and minor exporters.

## 3.1 THE PATTERN OF PRODUCTION AND CONSUMPTION

Tin is an almost archetypal UNCTAD 'core' commodity in the sense that its production and consumption are highly polarised between the developed and developing countries. Table 3.1 shows that approximately 77 per cent of world mine production comes from four developing countries - Malaysia, Bolivia, Thailand and Indonesia - and 61 per cent from South East Asia alone. These figures drop to 64 per cent and 51 per cent respectively if the (rather unreliable) statistics for the communist countries are added to the world total. Table 3.2 shows that developing countries had a higher market share in world tin exports (excluding those of China) than in any other non-fuel mineral of major interest to the Third World. Of the developed countries who produce tin, only Australia and South Africa are net exporters. China in the early 1970's exported about as much as Australia's production today, but its exports have greatly declined since than, and the USSR is a net importer (1).

Tin is regarded as one of the few minerals (along with silver) whose reserves are genuinely in short supply (Bosson and Varon, 1977, p.59). Table 3.2 shows that the largest producer, Malaysia, is depleting at a rate which would exhaust its known reserves in little more than a dozen years. According to US Bureau of Mines estimates, world reserves are 10 million tonnes, enough for 40 to 45 years current production and 'sub-marginal, hypothetical and speculative' deposits could allow another 100 years after that (Harris, 1978, p.5). Thailand's reserves appear to be substantially larger than those of Malaysia, and Indonesia's are three times the Malaysian figure, so considerable changes in the pattern of production could occur. Reserve figures, however, should be treated with extreme caution. Adding 'sub-marginal, speculation and hypothetical' deposits to the reserve figures would give Malaysia a larger reserve than Indonesia and nearly as large as Thailand (Harris, 1978, p.5). More important, all past estimates have been highly inaccurate. The US government's Paley Commission in 1952 estimated Malaysia's reserves at a figure less than the actual production since then (Page, 1977, p.5.5). The Fermor report on the mining industry of Malaysia in 1939 gave tin reserve figures for the country(2) of only 610,000 tons - less than the present day reserves (Fermor, 1939, p.119). Even long established producers like Malaysia, Indonesia and Thailand can by no means be said fully to have prospected their land area for tin, and prospecting in new producers like Brazil is only beginning. Moreover, the US Bureau of Mines estimates differ considerably from those of the ITC (see Engel and Allen, 1979, chapter one) and the former are used here in preference only because they are presented in a consistent form which allows comparison between countries. Much prospecting is done by private companies, too, and they are likely to be reluctant to disclose their true reserve figures.

'Reserves' anyway depend very much on price and technology. Page (1977, p.5.1) makes the observation that if tin were extracted from rocks containing as low a concentration as that at which platinum is currently mined, the top mile of the earth's crust would contain tin reserves enough for the next 30 million years! Of course since platinum sells at about five hundred times the price of tin, the market for tin might become somewhat restricted, to say the least, if the extraction costs were to be recouped from consumers. What does seem clear, however, in spite of the reservations about reserve figure, is that the dependence on Third World tin supplies will continue. Western Europe, North America and Australia combined have lower known reserves than Malaysia alone, and they generally have been more thoroughly prospected.

Tin has carried further than most metals the tendency for developing countries to process an increased proportion of their ore into metal. Malaysia, Thailand and Indonesia smelt virtually the whole of their output, as do Nigeria and Brazil. Bolivia is making strenuous efforts in that direction. Thus the bulk of world tin trade is in tin metal, whereas for minerals as a whole only about a third of LDC exports are processed fully into metal. In fact, according to 1970 statistics quoted in Bosson and Varon (1977, pp.87-89) a higher proportion of LDC tin exports were processed than was the case in any other major mineral, though copper followed closely and a high proportion of lead and nickel were processed too. The pattern common in other metals of major processing facilities being located in Western consuming countries and Japan is thus rare in tin, and indeed the largest purchaser of tin concentrates (unsmelted ore) is Malaysia (see Table 3.4). Malaysia's smelting dates back to before the First World War, and before Thailand set up its own smelter in the 1960's Malaysia smelted practically the entire Thai output too. Indonesia's self-sufficiency in smelting is very recent and previously a significant part of its output was processed in the Netherlands (and some in Malaysia). Bolivia's difficulties in achieving self-sufficiency in smelting, and the desirability of that aim, will be discussed in later chapters, along with the development of smelting in the other countries. Essentially they derive from the fact that her ores are complex and hard-rock whereas those in South East Asia and many other producers are alluvial. In alluvial tin mining ore is generally concentrated to about 70 to 75 per cent purity at the mine and smelting thus involves about a 25 per cent weight loss. Smelting costs are about 1 to 5 per cent of gross output value in South East Asia, which indicates it is a relatively minor undertaking beside, say, bauxite processing. In tin, too, smelting produces the virtually pure metal and there is normally no further refining stage.

The consumption statistics in Table 3.3 show that developing countries as a whole consumed less tin than Japan alone, and that 88 per cent of (non-communist) world tin produced is used in rich countries. In view of the dependence of rich countries on Third World tin production and the fact that US mine production is minimal (under 100 tons in 1978), the cartelisation potential of the industry is often discussed. Certainly the price elasticity of demand is low (3) at least in the short run. The established producers are small in number and few new ones (except perhaps Brazil) are likely to appear to break a cartel with significantly increased production.

In spite of this there are two main reasons why cartelisation is unlikely. First, the governments of the South East Asian producers have expressed themselves frequently against a 'confrontationist' approach to commodity negotiations and prefer to work through the International Tin Agreement (discussed below) on whose Council consumers are also represented. All three South East Asian countries are heavily private enterprise-orientated economies (especially Malaysia), keen (at least in principle) to continue to attract foreign investment, and committed politically to a pro-western stance. Bolivia, the most likely supporter of a cartel, is the highest cost and therefore most vulnerable producer. Bolivia also depends for about 57 per cent of her export earnings on tin, whereas the figures for Malaysia, Thailand and Indonesia are only 11 per cent, 7 per cent and 2 per cent, respectively (1977 figures).

The second and compelling reason is that the USA maintains a 200,000 tonne stockpile of tin built up for strategic purposes, equivalent in size to about a year's world production and officially nearly 170,000 tonnes in excess of strategic requirements (4). Stockpiling had started early in the Second World War, reaching 100,000 tons by 1942, but the major build up occurred in the 1950's, reaching probably about 350,000 tons in 1957 (5). This size, equivalent to seven years' US consumption, has been reduced by controlled releases administered by the US General Services Administration since 1962. By 1974 over 140,000 tonnes had been released, including 40,000 during the 1973-74 commodities boom (6). Although the USA has agreed with International Tin Council since 1966 to undertake stockpile disposals in such a way as to avoid market disruption, there is evidence from the 1950's - when the US imposed a buying strike on tin in an attempt to force down price - that aggressive retaliation in face of a cartel cannot be ruled out.

In the longer term, and leaving aside stockpile releases, cartelisation would depend for its success on the reaction of consumers to substantial price changes. In the absence of useful econometric evidence on the long term price elasticity of demand, some general points about consumption trends can be made. This leads to a consideration of the likely future course of prices. Since cartelisation is unlikely, market price trends are a good deal more relevant to producer policy.

Historically the consumption of tin has risen slowly, averaging growth of only 0.8 per cent per year from 1935 to 1972. Over the longer period from 1900 to 1972 tin metal consumption has risen 2.2

times, compared with 4 times for lead, 6 times for zinc and 12 times for copper (Fox, 1974, pp.17-18). Tinplate is still the largest use and accounts for 36 per cent of US and 43 per cent of British tin metal consumption (ITC, MSB, February 1980). The other major uses are in solder and in tin alloys and chemicals. It is well known that electrolytic tinplating has greatly reduced tin consumption per unit of tinplate compared with the older hot-dip process used before the Second World War. At present a food tin can uses only 20 per cent of the tin (and 65 per cent of the steel) used in 1945 (7). This reduction in tin per unit of final product has to some extent been counter-balanced by a growth in the use of cans in preference to other packaging materials, particularly glass. Over 40 per cent of cans used in the USA are now for beer and other beverages, though there are pressures from environmentalists against this use. This growing market is shared with aluminium cans, and tin-free steel (which is steel plated with chromium) is also a competitor with tinplate. The structure of the world tin industry (discussed further in the next section) does not predispose tin product users towards any vested interest in continuing to use tin. Tinplate producers are overwhelmingly makers of steel - the British Steel Corporation is the UK's largest tinplate manufacturer for instance - and tinfree steel has been developed by them, particularly the Japanese. Several factors give rise to expectations of growth in tinplate consumption *vis-à-vis* its rivals, however. Electrolytic tinplating itself has improved the competitive position of tin against other packaging. More recently, the post OPEC rises in power costs greatly favour tin - a tin can uses in its manufacture a third of the energy required by a comparable aluminium can, and 10 per cent of that needed for a returnable glass bottle. Innovations such as the two-piece can also enhance the attractiveness of cans relative to other containers since they are light, strong and more easily decorated than the older (three-piece) variety. During 1979 increases in world tinplate capacity outweighed the conversion of several US and Japanese plants to tin free steel. In spite of the low short run price elasticities of demand referred to earlier, price may well be important in the longer term. It is possible to install dual tinplate/tin free steel manufacturing facilities, and in the US can plants generally are built with convertibility between steel and aluminium. A detailed consultant's survey in 1978 (summarised in TI, January 1979) concluded that switches away from tinplate would occur only if the tin price rose above US $10 per lb at the 1978 general price level - which was 70 per cent above the average New York tin price for that year and nearly M$3,000 a picul at the then exchange rates.

In other uses too the competitiveness position of tin has not significantly weakened. Although the traditional uses of solder in such areas as plumbing have diminished, new growth has occurred particularly in electronics with printed circuit soldering. All in all these considerations lend support to a belief that the demand for tin is facing no immediate demise. Much further research would be necessary to predict future consumption trends (8) and particularly the growth of demand as LDC's establish more of their own canning capacity. In any case much uncertainty is involved, not least with regard to innovation. The US Bureau of Mines estimates (Harris, 1978, pp.13-15) suggest that over the period 1976 to 2000 world demand for tin metal will grow at the rate of 1.1 per cent *per annum* of which 1 per cent will be for primary tin. This is composed of a US demand growth rate of 0.4 per

TABLE 3.1
TIN PRODUCTION AND EXPORTS OF TEN LARGEST PRODUCERS, 1978
('000 tonnes)

| | TIN-IN-CONCENTRATES | TIN METAL | | MINE PRODUCTION AS PERCENTAGE OF RESERVES |
|---|---|---|---|---|
| | Production | Production | Export | |
| Malaysia | 62.6 | 71.9 | 70.1 | 7.5 |
| Bolivia | 30.9 | 16.2 | 12.5 (1977) | 3.1 |
| Thailand | 30.2 | 28.9 | 28.9 | 2.5 |
| Indonesia | 27.4 | 25.8 | 22.9 | 1.1 |
| Australia | 11.7 | 5.1 | 2.1 | 3.5 |
| Brazil | 6.5 | 8.3 | 1.7 | 1.1 |
| Zaire | 3.4 | 0.9 | 1.1 | 1.7 |
| South Africa | 2.9 | 1.5 | - | n.a. |
| Nigeria | 2.9 | 3.0 | 3.1 | 1.0 |
| UK | 2.8 | 8.4 | - | 1.1 |
| World (excluding USSR, China and some small socialist countries) | 197.3 | 192.3 | 167.6 | 2.5 |

Sources and Notes

(1) All production and export figures from ITC, MSB, February 1980.

(2) Tin-in-concentrates refers to the metal content of unsmelted concentrates (i.e. treated ore).

(3) China and the USSR are also significant producers. Chinese production figures of 20,000 tonnes of tin-in-concentrates and 20,000 tonnes of metal were quoted by World Metal Statistics, May 1979, which also gave figures of 18,000 and 17,500 tonnes, respectively, for the USSR. Neither set of figures is reliable. World mine production including the socialist countries is given as 235,100 tonnes. Chinese exports of over 3,400 tonnes were reported by importing countries in 1978 (TI, December 1979).

(4) In addition to primary tin metal production, there was recycled tin production of 9,000 tonnes, of which over half was from Europe.

(5) Reserve figures from Harris (1978) p.5.

TABLE 3.2

DEVELOPING COUNTRIES' SHARE OF WORLD EXPORTS OF TIN AND OTHER SELECTED MINERALS, 1975-77 AVERAGES

(Percentages of World Gross Exports)

| | | | |
|---|---|---|---|
| Tin | 82.6 | Manganese Ore | 49.7 |
| Bauxite | 78.4 | Iron Ore | 39.6 |
| Phosphate Rock | 63.2 | Lead | 25.1 |
| Copper | 56.9 | Zinc | 23.7 |

Sources and Notes

(1) From World Bank (1979). These are all the non-fuel minerals included in the Bank's list of thirty-three commodities of major interest to developing countries.

(2) World totals exclude centrally planned Asian economies, but include Eastern Europe and the USSR.

TABLE 3.3
CONSUMPTION OF TIN METAL, 1978
('000 tonnes)

| | | | |
|---|---|---|---|
| USA | 49.2 | Europe (excluding USSR) | 62.7 |
| Japan | 29.0 | Latin America | 9.8 |
| FR Germany | 15.0 | Asia (excluding Japan and China) | 10.7 |
| UK | 13.9 | Africa (excluding South Africa) | 1.8 |
| France | 10.3 | China | 16.0 (1977) |
| | | USSR | 25.0 (1977) |

Source
World Metal Statistics, May 1979.

TABLE 3.4
MAJOR IMPORTERS OF TIN-IN-CONCENTRATES,1978
('000 tonnes)

| | | | |
|---|---|---|---|
| Malaysia | 9.4 | Benelux | 3.3 |
| UK | 7.0 | Spain | 3.1 |
| FR Germany | 4.2 | Singapore | 2.3 |
| USA | 3.9 | Brazil | 2.3 |
| | | Netherlands | 1.5 |

Source ITC, MSB, February 1980.

cent and 1.3 per cent for the rest of the world. Recycling, according to ITC statistics, generated 21 per cent of 1978 US tin supplies, compared with only 13 per cent in the UK. Recycling is expensive since tin is a minor constituent of the final product from which it is to be reclaimed, but the Bureau of Mines statistics allow for a demand growth in recycled tin almost double that for metal. Overall, the growth of demand is estimated to be more than three times faster in the rest of the world than in the US.

Future supply trends depend on reserve availability and on the related trend in extraction costs as lower grade deposits are worked. Important too is the 'climate' for domestic and foreign investment in mining. These issues will be discussed in later chapters, but one may note here that if the trends of the 1970's continue, major increases in production seem unlikely. If the US stockpile surplus tin were released gradually over the period to 2000, this would add only 5 per cent a year to the world supply. Page, for example, (1977, chapter four) foresees a gradually rising real price trend as production grows by slightly less than consumption. This seems a sensible and conservative view to take as a working hypothesis, but it will be examined in the light of the investment appraisals to be presented in chapter six.

Price fluctuations, as well as the price level, are of concern to producers and consumers. Fluctuations in price in a predominantly demand-shift market like tin tend to lead to fluctuations in export earnings and producer incomes. While the effects on import availability of fluctuations in export earnings can in principle be counteracted at the cost of holding larger foreign exchange reserves, the associated fluctuations in government tax revenue from mining and in producer income cause planning difficulties. Further, in a country heavily dependent for export earnings on a particular commodity (as is Bolivia on tin) fluctuations in export income from it can lead to fluctuations in aggregate domestic demand. These, it is often claimed, are inflationary and inhibit investment. Price fluctuations (and the associated shortages and surpluses) also lead to increased costs for consumers, principally in the form of the need to hold increased stocks and to hedge on future markets, who may then prefer to pay a higher price for a substitute if the substitute's price is more stable.

Of the major tin producers only Bolivia appears to have an export earnings instability problem, according to a recent study by Hallwood (1977) and Bolivia's instability index (percentage annual deviations from a logarithmic trend) was only 8 per cent (9). For virtually all the other producers too, tin is too small a proportion of export earnings to have a significant effect on national income fluctuations anyway, so it is the effect on producing firms which is relevant. Here there is evidence that fluctuations may result in a bias towards capital intensive techniques, whose low prime costs allow mines to avoid the cash flow problems caused by low prices (10). For small mines with high operating costs, periods of low price are likely to mean closure, hence price fluctuations can lead to substantial fluctuations in supply.

The course of prices over the post war period has been influenced by US stockpile releases and, since 1956, by the operation of the International Tin Agreements. A great deal has been written about the ITA (11) and the main points can be made quite briefly. Tin price fluctuations, though substantial , appear to have been less than those in some other metals. Page (1977, p.12.2) shows that over the period 1950 to 1975, in about three-quarters of the years the average annual price change was greater in copper, zinc and lead. Within each year over the same period their fluctuations have been greater than those of tin in more than half the years in the case of zinc and two-thirds of the years in the cases of copper and lead. Note though that whereas the bulk of tin is traded through free world markets - the main physical market being Penang in Malaysia and the main 'paper' markets being in London and New York - part of world copper output has been sold at 'producer prices' fixed for long periods. The free market price is then used to clear residual supplies and thus is less stable. Producer prices also have been used in zinc (Page, 1977, pp.12.6-12.7) (12).

The extent to which the ITA and US stockpile disposals have influenced market prices and producer revenue has been estimated by Smith and Schink (1976) for the period 1956 to 1973. The GSA disposals are likely to smooth fluctuations in the sense that releases have been greater then prices are high (to avoid 'market disruption' as well as to make profit). The ITA influences the market by purchases and sales of tin from its buffer stock, and by its imposition of export control during periods of sustained low prices. Floor and ceiling prices are set by agreement in the International Tin Council, which runs the ITA, and are subject to periodic revision. The buffer stock market manager has discretion whether to buy within the lower of three price bands, to buy or sell within the middle band, or to sell within the top band; but when the price reaches the 'floor' at the bottom of the lower band he must buy, and similarly he must sell at the 'ceiling'. Export control to support the floor is effectively production control, since producers are allowed only a limited stock build up.

Smith and Schink's simulations suggest that average annual changes in price (prices deflated by the UN index of dollar price of exports by LDC's, deflated by changes in the pound/dollar exchange rate) in the absence of ITA and GSA activity would have been 10.4 per cent. The ITA reduced them only to 9.4 per cent (though the model cannot simulate ITA production control) whereas GSA activity reduced them to 6.9 per cent. The GSA operations reduced (discounted) producer income (13) by about 5 per cent over the period, presumably by creaming off the top of the market. The possibility (well known to the commodity control literature) that the ITA itself might have reduced producer income did not occur (14). Also, according to Hallwood's (1979) estimates (in the absence of full published accounts for the ITA) the tin agreements have been broadly self-financing over the period 1956 to 1977. The tin agreement has been the most durable of all post-war commodity agreements and a sixth ITA is to be negotiated for 1981. While the agreements have been extremely successful in holding the floor price, the ceiling has been breached many times and it is in this regard that the GSA releases have been a factor of much more significance in stabilising the market. Smith and Schink's consequent conclusion is that the ITA's longevity is due to its ineffectiveness, since consumers could rely on the GSA both to protect breaches of the ceiling and to take action if producers attempted to force tbrough a large rise in the floor price (1977, p.721). Thus serious confrontation did not arise.

This rather harsh judgement on the ITA has been substantiated to some extent by events during the current (Fifth) agreement (1976 to 1981). The buffer stock has been unable to contain the ceiling and has been out of tin since January 1977, the stock having stood at its maximum, normal level of 20,000 tonnes (15) only a year previously. The buffer stock size is still small compared with the GSA stockpile, and over the period 1966 to 1974 Smith and Schink have estimated that stock of 40,000 to 70,000 tons would have had to be maintained (1976, p.725) in the absence of stockpile releases to contain prices within alternative ranges of 5 per cent, 10 per cent and 15 per cent around the (logarithmic) price trend (16).

As earlier commentators have noted, the ITA price range has tended to lag behind market prices, so the agreement has only had to defend a very low floor whereas a high but unspecified ceiling was defended by the stockpile (Smith and Schink, 1976, p.720; and Fox, 1974, pp.391-93). Considerable producer dissatisfaction has led to changes. Under the fifth agreement the price range is now reviewed by an Economic and Price Review Panel every six months, which puts proposals to the full International Tin Council, and explicitly takes account of production costs and returns to investment. The EPRP reviews costs regularly and in 1977 commissioned a series of investment appraisals to determine the prices needed to secure new production (see chapter six). The ITA price range of M$1,500-1,950 (17) agreed in July 1979 had a floor price above the 1978 costs of the small scale Malaysian gravel pump mines, and the Bolivian underground mine costs were nearer the ceiling (Engel and Allen, 1979, p.127). A new range of M$1,650-2,145 agreed from March 1980 still has a <u>ceiling</u> hardly higher than Bolivia's expected costs (<u>TI</u>, April 1980). However these costs include export duties. Export duties are indeed a cost to the mines but they are in the power of producer governments to reduce. Bolivia though is a special case among producers operating very different deposits from most others (see sections 2 and 3) and special measures were necessary to secure her joining the fifth agreement, principally the signing of the so-called La Paz Charter in February 1977 (between Bolivia and the three South East Asian producers). The charter recognised Bolivia's position as a high cost producer and this was instrumental in the formation of the EPRP (<u>TI</u>, February 1980).

In 1979 and 1980 more belligerent positions were taken by the ITA participants, partly as a preliminary to negotiations (in March and April 1980) for the sixth agreement (to come into force in July 1981) and partly with regard to stockpile releases agreed by the US Congress at the end of 1979. Since 1977 many operators in the market have described it as being in 'surplus', in the sense of Western consumption growing less rapidly than non-communist world production, and hostility has been expressed by normally quite moderate spokesmen. Thus the Malaysian Minister of Primary Industries and the Director of Marketing at the Indonesian State Tin Company have both threatened withdrawal from the ITA over stockpile releases with the possibility of an OPEC style organisation being established. The idea of producers buying up stockpile releases to hold as assets has also been mooted, and annual ministerial meetings of the tin producers in the ITA have started. The US has been making belligerent noises too, having joined the fifth agreement without previously having been a member. The US suggests that the next ITA should operate a buffer stock without production control, and has been reported as saying that the price range should not be based necessarily on costs (TI July, August, September 1979). Whether these positions will be toned down in practice remains to be seen. It is certainly worth noting with regard to the general 'success' of the ITA that a well defended and realistic floor price, regularly raised to allow for inflation, is invaluable not only for investment planning but also to avoid the supply fluctuations associated with the entry and exit of small marginal mines in South East Asia in response to price changes.

## 3.2 THE TECHNOLOGY AND STRUCTURE OF THE INDUSTRY

In earlier chapters the importance was stressed of a bargaining element between foreign companies and host-countries in the distribution of benefits from mineral exports and investment. Important too is the interaction between local mining companies and foreign multinationals. In an influential work on US multinational companies Vernon (1973, p.35) notes that the variation in the degree of MNC involvement as between different commodity exports is by no means accidental. He attributes the variations to differences in economies of scale, in the complexity of the production technology, and to the importance of captive markets overseas as an outlet for raw material exports. Often these differences are reflected in the degree of vertical integration of an industry, and in highly vertically integrated activities MNC's in principle can exert great market power both in bargaining with host governments and in excluding local companies.

Among the (non-oil) minerals bauxite is usually cited as an extreme case of MNC market power. Bauxite reserves are found quite widely, so MNC's can switch their source of supply if particular host governments are recalcitrant. At the same time the initial capital costs and high technology of extraction, and more particularly of refining and smelting, make entry difficult for private local firms, while state companies can find it hard to sell their output (whether of bauxite, alumina or aluminium) in a world market dominated by the oligopolistic aluminium companies. In contrast to this situation the tin industry is often described as 'unstructured' or 'fragmented' (18) and Fox (1974, pp.6-8) notes the generally smaller scale of the industry in relation to other non-ferrous metals. The next chapter will explore the relation between the development of technology and the entry of multinational companies. For the moment, the available technologies are considered with a view to seeing whether they seem to predispose the tin industry towards a particular kind of structure.(19)

The most important mining technology in terms of world output is gravel pumping (see Table 3.5) whose individual units generally work very small deposits, though multiple gravel pump operations on larger deposits are possible (20). Gravel pumping, and all the other techniques except underground mining (21), are used to work alluvial deposits (where tin ore has been broken down by the action of water and occurs fairly near the surface) or eluvial deposits (where the primary tin deposit has also been decomposed but, unlike the alluvial deposit, it has not been transported any distance by water).

Gravel pumping takes place in an open pit and the tin ore is mined from its sides by being broken down by high speed water jets directed by rigid hosepipes ('monitors'). The liquid ore and mud is sucked up from the bottom of the pit by pumps ('gravel pumps') and elevated on to a sluice for primary concentration. Further concentration can be undertaken by jigs, and in some cases (e.g. for very fine grain ores) jigs are used for primary treatment. A gravel pump mine can operate to a depth of about 150 feet, with double pumping, though they are usually much shallower.

Dredges operate on a pond (or 'paddock') largely of their own making, or on the sea, and are ship-like structures which dig the ore from the bottom with a continuous bucket chain. The paydirt is concentrated on the dredge using jigs. Dredges can work to a maximum depth of about 150 feet too, and inland dredges can dig deeper if the level of their paddock water is lowered. Though bucket dredges are the most common by far, sometimes suction dredges are used which suck ore from the bottom through a pipe, generally nowadays attached to a revolving cutter.

Underground mining for tin is basically similar to the mining of coal in Europe and is used to mine primary ('hard rock') deposits.

Suction boats operate only off the southern coast of Thailand and are of recent origin. They are miniature suction dredges, usually with the suction pipe operated by a diver. Originally, in the early 1970's, converted fishing boats were used, though now purpose built boats are available which can dig to a maximum depth of about 60 feet.

Dulang washing is the working of surface tin deposits by panning, rather like early gold mining in rivers in the USA. Dulang washers often work in teams and are usually women.

Open cast mining for tin is dry-mining, where no use is made of hydraulic methods. Excavators, draglines etc. are used to mine richer, deeper and more concentrated deposits.

The difference in ease of extraction between alluvial and hard-rock deposits is the basis of the considerable difference in costs between the South East Asian producers and Bolivia noted in the previous section. Underground mining is important too in Australia and the UK's mines are exclusively underground, but in the Bolivian case, the deposits as well as being in deep lodes are situated in mountainous and inaccessible areas. Besides this cost difference and the large resource rents generated in South East Asia, Zaire and Nigeria, the other striking feature of the industry is the range of techniques used in each country (see Table 3.6). Even in Bolivia, underground mining of the shallower deposits can be done on a small scale, and in virtually all the other producers several distinct production techniques are available. To some extent this is the result of the widespread existence of small deposits in these countries which allows small scale methods to survive. This is true of minor techniques such as dulang washing and of the open cast mining of very deep alluvial deposits. More important, and among the main techniques, gravel pumping is generally an alternative to dredging inland (22); and for shallower deposits which are close inshore, suction boats can substitute for offshore dredges (23). Thus small scale technology has existed as a vehicle for the entry of local mining firms; and the competing techniques of offshore dredging and small boat mining, and gravel pumping and inland dredging, account for the bulk of South East Asian output. At present almost all gravel pump production in Malaysia and Thailand is under the control of local firms, as are many of the smaller dredging operations in Thailand. In Malaysia the Malaysia Mining Corporation, a federal organisation formed in the 1970's, operates major dredging concessions previously run by tin mining's largest multinational, the London Tin Corporation. MMC works in partnership with the international mining house Charter Consolidated, and the rest of the dredging sector is still mostly foreign controlled. In Indonesia nearly 90 per cent of output is produced by the State tin mining company, P. T. Timah, and the rest from three foreign companies. In Bolivia about three-quarters of tin output comes from the State corporation, Comibol.

TABLE 3.5
TIN MINING METHODS, 1979

| | PERCENTAGE OF (NON-COMMUNIST) WORLD OUTPUT | NUMBER OF UNITS OPERATING | TONNES OF TIN-IN-CONCENTRATES PER UNIT PER YEAR |
|---|---|---|---|
| Gravel pumps | 28.5 | 1304 | 44 |
| Underground mines | 22.0 | 33 | 1333 |
| Dredges - onshore | 13.2 | 90 | 295 |
| - offshore | 6.5 | 24 | 547 |
| Suction boats | 7.5 | c.2000 | 7 |
| Dulang washing | 1.9 | c.27500 (workers) | 0.1 |
| Opencast mines | 1.7 | 23 | 152 |
| Minor methods not reported to ITC | 18.7 | - | - |

Sources and Notes

(1) Figures from ITC, MSB, February 1980.

(2) Covers the period January to June only. This may marginally overstate the share of suction boats, which in 1978 produced 54 per cent of their output in the first six months.

(3) The ITC statistics estimate 4500 dulang washers in Thailand, with no estimate for Malaysia. In Malaysia in 1975 there were nearly 23,000 licensed dulang washers and this figure has been added to the total. See Department of Mines Malaysia, BSMI, 1975. Similarly, no estimate is given for the number of suction boats, but a figure of 2139 is quoted in the same source for 1979 for Thailand, the only country where they operate.

TABLE 3.6
TIN MINING METHODS IN MAJOR PRODUCING COUNTRIES
(percentages of total output)

| MALAYSIA | 1928 | 1960 | 1974 | 1979 (second quarter) | 1979 (number of units) |
|---|---|---|---|---|---|
| Dredging | 30 | 54 | 32 | 32 | 54 |
| Gravel Pumping | 45 | 34 | 53 | 54 | 801 |
| Open Cast | 6 | 2 | 4 | 4 | 21 |
| Underground | 6 | 4 | 3 | 2 | 22 |
| Dulang | 2 | 2 | 4 | 5 | - |
| BOLIVIA | 1973 | 1976 | 1978 (figures for selected quarters) | | |
| Comibol | 68 | 71 | 73 | | - |
| Medium mines | 19 | 11 | 10 | | - |
| Small mines | 11 | 15 | 17 | | - |
| INDONESIA | 1954 | 1960 | 1978 | | 1979 (number of units) |
| Dredging | 68 | 66 | 34 Offshore<br>16 Onshore | | 15<br>21 |
| Gravel Pumping | 30 | 31 | 50 | | 177 |
| THAILAND | 1951 | 1960 | 1978 | | 1979 (number of units) |
| Dredging | 58 | 48 | 17 | | 24 |
| Suction boats | - | - | 34 | | 2139 |
| Gravel Pumping | 24 | 35 | 32 | | 355 |
| Other | 18 | 17 | 17 | | 303 |

Sources

ITC, MSB, October 1977 and February 1980; Thoburn (1977b, p.87) for Malaysia 1928-74; Robertson (1964, p.63) for Indonesia 1954 and 1960, and Wit (1971, p.46) for Thailand 1951 and 1960.

The existence of successful small scale technology in tin, though helped by the availability of many deposits too small for the methods favoured by foreign companies, also owes much (as chapter four will show) to successful experimentation by local miners. Chapter four will show too that there has been some local private sector entry into dredging. Underground mining in Bolivia owed much of its development, prior to nationalisation, to the Patiño organisation, a locally originating multinational, and in the other LDC producers it is not of sufficient importance to merit special discussion at this stage. In dredging it does not appear that the technology is in itself too complex or too difficult to obtain. As the experience of recently formed local Asian dredging companies has shown, management can be hired on the open market though it is often still expatriate. Nor is design technology monopolised by the mining multinationals. Indeed, it is more usually bought from consultants - the main ones being Paynes in England, a specialist dredge design firm, and Mining and Transport Engineering of Amsterdam, a subsidiary of the Dutch shipbuilding firm IHC. Further, there exist in Malaysia engineering firms capable of fabricating and erecting dredges, who have also worked in Thailand, and who are predominantly local Chinese (like the miners). Both the existence of the design consultants and of local engineering firms dates to before the Second World War (24).

The vertical structure of the industry also has not been a barrier to local mining companies entering tin. Taking the metal's main use in packaging there are four stages beyond mining, namely smelting, marketing, tinplate manufacture and can manufacture. MNC control over the smelting and marketing of Bolivia's production has weakened the bargaining power of that country's state tin corporation and the government has taken measures to establish local smelting. In the case of most other producers smelting (of alluvial ores) is subject to large economies of scale, but the technology has not been too complex in the past to prevent one large smelter in Malaysia being formed by local interests. The degree of vertical integration between mining companies and smelting companies has not been great and independently owned mines and smelters have not been disadvantaged greatly by their lack of vertical links. Similarly most marketing has been done on an arm's length basis. One technological explanation of this, as noted earlier, is that smelting is a small part of the value of output and generally no further refining is necessary.

Also important is the fact that there are very few links between smelting and mining companies on the one hand and tinplate manufacturers on the other. Tinplate is overwhelmingly produced by steel companies and tin is a minor part of total costs. The tinplate manufacturing industry is quite concentrated, however. In 1971 for instance the single UK manufacturer bought 8,000 tons of tin, the four manufacturers in Japan 14,000 tons in total, and the seven in the USA bought 24,000 tons between them (Hoare, in ITC, 1974, p.31). This could give to individual manufacturers fearing for their supply sources and wishing to integrate backwards to secure them. In fact, the tendency for tin to be sold on the open market has greatly alleviated any potential supply insecurity. Moreover, historical processes have tended to cumulate and the lack of backward integration by some companies has meant that the others have less incentive to move backwards either. Long term contracts too, such as those between Japan and Indonesia, reduce supply insecurity. The high concentration among tinplate manufacturers does give rise to a potentially strong bargaining position, but there seems to be little hard evidence of their behaving collusively. Tin can manufacture is even more concentrated, five MNC's dominating the market (25) (TI, September 1978). This industry structure is likely to make for difficulties for producers moving into downstream activities, and tinplate manufacture will depend much more on their being a domestic steel industry than on domestic tin production.

In mining, therefore, the structure of the world market for tin metal does not seem to be a factor enhancing the power of MNC's relative to local companies. More important a barrier to local entry has been the high capital costs of dredging relative to the smaller scale techniques, in a situation where local Asian (and African) producers normally use partnerships or private companies rather than public limited companies. In Malaysia or Thailand a gravel pump mine could be established for about US $500,000 whereas a large inland dredge project would cost about US $15 million (see Table 6.1). Offshore projects would be more expensive still, a large one in Indonesia costing nearly US $35 million; whereas a Thai suction boat costs only about US $75,000. In Bolivia, an underground mine would take about US $13 million to establish (Engel and Allen, 1979, p.145). It is important to emphasise though that by world mining standards these dredge and underground mining investment costs do not make tin mining a large scale operation. According to Bosson and Varon's (1977, p.46) study, based on 1973 data, 75 per cent of the total investment expenditure in a large sample of mining ventures was in projects costing over US $100 million, and inflation since 1973 would have increased those costs considerably. In other words, capital costs have been high enough to deter local people from investing in certain mining techniques but the really large investments needed in other metals have not been in evidence. In consequence the services of large international mining houses have not been required to develop the industry in the past (Fox, 1974, p.8); and the importance of specialist tin companies, exemplified by the London Tin Corporation, is understandable to some extent in terms of capital requirements. The fact that most tin is mined by alluvial methods, whereas dry, open cast mining is increasingly used in other metals has also been a factor making for some separateness in terms of technical expertise.

In the world mining industry considerable technological change has occurred with consequent alternations in industry structure as larger scales of operation have been developed to deal with high throughput, low grade operations. Increasing capital requirements have also bought the oil companies into mining through takeovers. Tin has experienced some of these changes and the grades operated have certainly declined in the main producers. In the same way that the size of trucks and mechanical shovels used in open cast hard-rock mining has increased several-fold over the last twenty years (Bosson and Varon, 1977, p.34) the throughput of mines has increased in tin (26). Before the Second World War dredges typically used 7-14 cubic feet buckets and a monthly throughput of 200,000 cubic yards was quite large. Since the 1960's 600,000 cubic yards has become common for new dredges and bucket size has increased to over 20 cubic feet. In the late 1970's further increases in size have taken place and one dredge commissioned at the end of 1979 had 30 cubic feet buckets (27). Computerisation too has affected tin as well as other metals and the newest offshore dredges use computer control. One of the most advanced and the largest dredge to date (Billiton's offshore dredge in Indonesia) thus has been part of a project costing US $65 million of which US $30 million was for the dredge alone. This is still small compared with other metals, but it is thought that a huge new deposit found in Malaysia at Kuala Langat in Selangor state will be worked by a deep multiple dredging project costing over US $100 million. Financing methods too have developed to meet these larger capital requirements including the use of leasing from banks. Two oil companies, Shell and British Petroleum, now have interests in companies associated with tin.

In underground mining too some changes in techniques have occurred, particularly block caving in Bolivia, again to allow the mining of lower grades of ore. Equally important, the locally owned sectors of the tin industry have shown adaptability to lower grades. Gravel pump mines have increased their throughputs, particularly in Malaysia where low grades are more usual than in Indonesia or Thailand. Earth moving equipment is used to drystrip and to feed the monitors and more mechanised forms of ore treatment are used. The Thai suction boats can hardly be seen as a response to low grades - they work deposits over twenty times richer than those of the Malaysian gravel pump miners - but they are certainly an interesting innovation which has dramatically altered the balance of economic power in the industry. Attempts to scale down dredges, so as to work small deposits with less labour than needed for a gravel pump operation have been tried. Experiments were conducted in Brazil, but when tried in Indonesia in the 1960's small dredges which could be dismantled were a failure (28). More work is being conducted on them in Malaysia at present (<u>TI</u>, January 1980), so their failure is not yet certain.

The availability of small scale technology and the size and complexity of that used in larger operations no doubt influences industry structure, and interacts with it. This section should end with a note of caution, however, and McKern's (1976) work on Australian mining is instructive in this regard. His study of a dozen minerals compares their domestic to their international market structure. Using stepwise multiple linear regressions he finds no general significant association between the degree of local ownership and any of his bargaining-strength variables: industrial concentration, technological intensity, size of capital investment, size and richness of deposits, and relative abundance of the mineral. Unfortunately he provides no detailed account of the Australian tin industry, though information is given on tin at various points. Tin is described as using a technology of 'low' complexity, with a 'medium'-complexity processing technology, and on his scheme it would be one where foreign entry was the result of horizontal diversification rather than vertical integration. Interestingly, though, in 1971 to 1972 tin had the lowest share of value added under foreign control of eight minerals for which data were available (29).

## 3.3 THE MAJOR EXPORTERS

Table 3.7 sets out the main economic characteristics of Malaysia, Bolivia, Thailand and Indonesia. The three South East Asian countries all lie on the same great tin belt, associated with granitic mountain ranges, which runs from Burma in the north to Indonesia in the south.

TABLE 3.7
ECONOMIC CHARACTERISTICS OF THE MAJOR TIN EXPORTERS, 1978

| | MALAYSIA | BOLIVIA | THAILAND | INDONESIA |
|---|---|---|---|---|
| Population (millions) | 13.3 | 5.3 | 44.5 | 136.0 |
| Land area (thousand square kilometres) | 330 | 1099 | 514 | 2027 |
| Gross National Product per head | $1090 | $510 | $490 | $360 |
| Average Annual Growth GNP per head 1960-78 | 3.9% | 2.2% | 4.6% | 4.1% |
| Average Annual Population Growth Rate 1970-78 | 2.7% | 2.6% | 2.7% | 1.8% |
| Average Annual Inflation Rate 1970-78 | 7.2% | 22.7% | 9.1% | 20.0% |
| Shares in GDP of - Agriculture | 25% | 17% | 27% | 31% |
| - Manufacturing | 17% | 13% | 18% | 9% |
| - Mining and Quarrying (1977) | 5% | 11% (1975) | 2% | 19% |
| Share of Exports in GDP | 51% | 17% | 21% | 21% |
| Share in (1977) Exports of | | | | |
| - Manufactures | 17% | 4% | 19% | 2% |
| - Agricultural products | 56% | 17% | 72% | 27% |
| - Fuels and minerals | 27% | 79% | 9% | 71% |
| (of which tin) | 11% | 57% | 7% | 2% |
| Workforce - Agriculture | 44% | 44% | 63% (1975) | 64% (1971) |
| - Manufacturing | 13% | 9% | 11% (1975) | 6% (1971) |
| - Mining & Quarrying | 2% | 6% | 0.4% (1975) | 0.2% (1971) |

Sources and Notes:

(1) Mining shares in GDP (in current prices except for Malaysia) from United Nations (1978b). Tin share in countries' exports from United Nations (1978a), with Bolivian tin exports for 1977 estimated from TI. Workforce shares for Bolivia & Thailand from ILO (1979). (This source does not list Malaysia or Indonesia (nor Brazil, Nigeria or Zaire). Malaysia from Malaysian Ministry of Finance Economic Report 1979-80 and Indonesia from Biro Pusat Statistik, Indikator Ekonomi June 1978.

(2) Other statistics from World Bank (1980b).

Though some primary tin deposits are to be found, erosion generally has broken down the tin into alluvial (and eluvial) deposits. The Bolivian deposits are primary ones and are not associated with the alluvial deposits of the other Latin American producer, Brazil.

3.3.1 Malaysia (30)

Malaysian growth has been more closely associated with primary commodity exports than that either of Thailand or Indonesia. It is an unusually open economy and has been so since the great expansion of tin exports in the late nineteenth century and the growth of the rubber industry in the early years of the twentieth. Rapid growth in oil palm and timber exports in the post-war period has lessened dependence on the traditional export staples somewhat, though Malaysia is still the world's largest exporter of natural rubber. The country was under the control of the British from 1874 to 1957, and British firms have accounted for the bulk of foreign ownership of the mines and plantations. Chinese immigrated into Malaya in large numbers and are now about a third of the total population, owning most of the rest of the mining sector and being important in trade, small industry and small rubber estates. Indian workers were imported to work on the large rubber estates and they form about 10 per cent of the total population. The indigenous Malays, or 'bumiputras', traditionally have formed the bulk of the rural population, among whom rice growing was the predominant activity, but Malays also own over half of the acreage of rubber smallholdings (defined as holdings under 100 acreas).

The highly multiracial nature of Malaysian society has led to some distinctive problems. Serious racial riots in 1969 broke a generally accepted implicit understanding that Malays would hold the political power and Chinese the economic power. As a direct result, official policy now stresses the 'restructuring' of asset ownership and employment in favour of the Malays, though there is also a declared aim to eradicate poverty regardless of race.

With a high rate of population and labour force growth, Malaysia faces the need to create employment. This is also an effective means of redistributing income and alleviating poverty. Malaysia is fortunate in having large quantities of unused land suitable for agriculture, roughly five million acres in Peninsular Malaysia and another ten million in East Malaysia (Sabah and Sarawak on the island of Borneo), which is in total almost double the present agricultural acreage. The state has taken a large and direct role in new land development, mainly for rubber and palm oil. Besides measures to alleviate poverty by in situ development such as further rubber replanting and irrigation schemes for rice, manufacturing is playing an increasing role in Malaysian development policy with (like Thailand) a growing orientation towards export. In the first two years of the (current) Third Malaysia Plan, agriculture had only 12 per cent of new job creation whereas manufacturing had 29 per cent (trade 16 per cent and government 21 per cent). Manufacturing is targeted to grow to 35 per cent of GDP by 1990. Foreign investment is strongly encouraged in manufacturing, though not in the primary industries. One reason seems to be that the restructuring of the workforce and the reserving of equity for Malays is easier when businesses are foreign owned than when they are controlled by local Chinese. Malaysia certainly hardly needs foreign investment for balance of payments reasons, its current account historically having been in surplus, and having been helped by new commodity exports and by high commodity prices in the 1970's. Malaysia is also a small net exporter of oil, and it has one of the highest *per capita* incomes in Asia. Like Indonesia and Thailand, it is a member of the regional grouping ASEAN - the Association of South East Asian Nations. ASEAN, however, has little in the nature of common policies or development strategy (31).

Tin is mined primarily in the two states of Perak and Selangor on the west coast, in the centre of more general economic activity. Mining other than tin (and petroleum) is unimportant.

### 3.3.2 Bolivia (32)

In the middle ages Bolivia was an important part of the Inca empire, and was conquered subsequently by the Spanish, for whom it was a major supplier of silver. The country has been independent since 1825, during which period it has had an unusually large number of changes of government. Mining, firstly of silver, then of tin, and more recently also of petroleum, has been a central feature of Bolivian development and, as later chapters will show, the important change of government in 1952, when the nationalist revolutionary movement took power, was closely associated with the power of the tin mine workers. At present, the mine workers are one of the main sources of opposition to the military junta which took over in November 1979.

Bolivia has a land area more than half that of Indonesia, yet has only 4 per cent of Indonesia's population. Compared to Malaysia and Thailand, too, Bolivia is very sparsely populated. The main area of population (about three-quarters of the total) is the high plateau area or *altiplano* and the eastern *cordillera* of the Andes. From this mountainous area a series of high valleys lead down to the low lying eastern region. Most of the tin and other metals is found in the mountainous areas in and around the *altiplano*, and the great height of the mines as well as their workings being underground adds to costs. Climate varies from subtropical in the lowlands to the very extreme cold of the mountains in winter. Oil is found in the lowlands, where there is also considerable potential for agricultural development, though this has been hampered in the past by lack of infrastructure. The continued role of minerals is illustrated by the low share of agriculture and manufacturing in exports and GDP, though agriculture still employs the bulk of the workforce.

Bolivia has been landlocked since the late nineteenth century when it lost its coastal region in a war with Chile. It also lost territory in the south east following war with Paraguay in the 1930's. Communications in the past have tended to run from the mountains to the Chilean coast, being concerned with the export of metals.

Current UN estimates of income per head put Bolivia ahead of Thailand, but these statistics have been disputed by the Bolivians (33).

Official government objectives, at least prior to the 1979 military takeover, included the reduction of personal and region income inequality and the creation of employment (Gillis, 1978, p.17). The income distributional problem is made more acute by the fact that roughly half of Bolivia's population is made up of indigenous Indians, the highest proportion in Latin America, who historically have been disadvantaged in a variety of ways.

Under the five year National Plan (1976-80) drawn up by the recently deposed government, the growth target was set at a level which has far exceeded actual achievements (34). Agriculture, though it has developed through land reform, cooperatisation, and new settlement since the 1950's in most recent years has lagged in output behind the rate of population growth. Tin mining production, as later chapters will show, has been deprived of funds for expansion, and petroleum exports have declined considerably as the result of a lack of exploration and investment in the past. Domestic consumption of oil rose from 41 per cent of production in 1976 to 75 per cent in 1978. The combination of stagnant mineral exports and rising food imports led to a sharp rise in the current account deficit over the 1976-78 period. By 1978 debt servicing was taking up almost half of export earnings (World Bank, 1980b, p.134) and the long term capital inflow was 30 per cent less than the current account deficit. A 25 per cent devaluation of the peso was carried out in November 1979.

Foreign investment is officially encouraged, even in the largely state controlled mining sector, and is admitted on terms broadly similar to those of other members of the Andean Pact (35). With certain exceptions foreign companies are expected to have majority Bolivian ownership within a twenty year period and incentives include tax holidays and exemption from import duties on equipment.

It is difficult to be optimistic about Bolivia's immediate economic prospects and the difficulties experienced by its tin industry (see chapters four and five) have not been counterbalanced by really successful development in other sectors.

### 3.3.3 Thailand (36)

Thailand is unusual among developing countries in never having been a western colony, having formed instead a buffer between British Malaya to the south and French Indochina to the north. The process of export expansion and colonial rule in Malaya was a powerful influence in the Thai economy nevertheless. Free trade was imposed on Thailand by the Bowring treaty of 1856 with Britain, and rice exports grew rapidly to a third of the world total before the Second World War, Malaya being a major customer. Other exports grew too, particularly rubber, tin and teak, but by the late 1920's rice was still about two-thirds of the total; and during the late nineteenth century and for most of the interwar period tin was about 10-15 per cent (Ingram, 1971, p.94). Tin thus never occupied the dominant position it held in Malaya before the coming of rubber, though, as in Malaya, it was an important factor in inducing Chinese immigration. Chinese are now some 10 per cent of the population, though they are assimilated to a much greater degree in Thai society than has occurred in Malaysia or Indonesia.

Thailand is a constitutional monarchy, royal power having been limited by a military coup in 1932. Apart from a brief period of civilian government in the 1970's, the military in practice have been the dominant element in the government since the 1930's.

The key position of domestic and export agriculture has continued well after the Second World War and, apart from a brief and unsuccessful period of direct government investment in industry in the 1950's, industrialisation began in the 1960's with an import substituting programme. Since the 1960's, Thailand has achieved an unusually high rate of growth based mainly on further increases in agricultural output. This agricultural growth predominantly has taken the form of an expansion in the cultivated area, though there has been a switching out of rice into newer cash crops such as maize or cassava, particularly by farmers in the central region. Thailand is still, however, a major exporter of rice, second only to the United States. Unlike Malaysia where new land development has relied heavily on state initiative, the Thai government has confined its role in agriculture mainly to providing roads into new areas and irrigation. Manufacturing development has begun to be more export orientated aided over the 1970's by the de facto depreciation of the Thai baht which drifted downwards with the US dollar to which it was tied until 1978 (it is now tied to a basket of currencies). Foreign investment is officially encouraged, but of the capital of the firms granted promotion certificates over the 1959-69 period, about two-thirds were Thai and much of the foreign capital was in majority Thai owned joint ventures (Ingram, 1971, pp.290-92).

The move towards export manufacturing has been more employment creating than was the import substitution programme, and this, combined with the agricultural growth, has meant a quite widespread participation among the population in the growth of income. Thus the World Bank estimates that the proportion of people in absolute poverty has falled from half to a quarter since the early 1960's. Nevertheless many farm households, especially in the north and north east, have not gained greatly, and this has been a factor in keeping down unskilled wage rates, given the high mobility between occupations in Thailand. The Bank's predictions for future economic performance are less optimistic than recent achievements would suggest. Principally this is because the Thai economy will run out of new land by the end of the 1980's if agricultural expansion continues, and the level of agricultural productivity will need to rise if growth is to be maintained. The Bank mission was also critical of the structure of tariffs, which generate a bias against local capital goods production, since effective protection rates are much higher on final consumer goods and investment promotion includes tariff rebates on capital goods imports. The tax system was also sharply criticised, both on equity and efficiency grounds, and there is a continuing and serious problem of tax evasion. The Thai government has explicitly recognised income redistribution as an objective along with economic growth since the third five year plan (1971-76) and this emphasis has continued in the present (fourth) plan (1977-81).

Mining in Thailand is far less important than in most of the other tin producers (37), and it is dominated by tin. Over 80 per cent of tin comes from the southern provinces, particularly Takuapa, Phanggna, Ranong and Phuket. This concentration is largely the result of geology, but historically mining has been restricted to the south by government decree.

### 3.3.4 Indonesia (38)

Indonesia achieved independence from the Dutch after the Second World War and the occupation by the Japanese. In terms of population it is the fifth largest country in the world, combining exceptionally high population density in the main island of Java (where live almost two-thirds of the country's population) with considerable unused land in its numerous outer islands. After independence, under President Sukarno's 'guided economy' the country reached a peak of dislocation in the mid 1960's accumulating high external debt and experiencing an inflation rate which reached 650 per cent. Many sectors were run down and <u>per capita</u> income growth was reversed. The new military regime which took over in 1966 placed a heavy stress on stabilisation, and has been in power ever since.

Table 3.7 shows clearly the importance to Indonesia of fuels and minerals. Petroleum alone is more than half of export earnings and of government revenue, and although it was produced in Indonesia even before the First World War, its dominance is recent and owes much to the OPEC price rises of the early 1970's. The impact of oil on a <u>per capita</u> basis is small, however, and its employment generation is minute. Part of the benefits of oil, anyway, were dissipated by unwise investments in the early 1970's by the state oil company, Pertamina. The bulk of the population is involved in growing rice and in export agriculture, particularly rubber. The supply of food and the creation of employment are probably the country's two most serious problems. Although rice production and productivity has increased with high yielding varieties, these have proved prone to disease, and population growth has increased demand to the extent that Indonesia purchased in 1977 one-third of the world imports of rice. The rubber industry has stagnated, and commentators place doubt on the medium term potential of the oil industry to expand exports in the face of growing domestic consumption of oil (whose sale is subsidised) and lack of recent exploration activity. Exports of palm oil, timber and new minerals such as copper and nickel have risen rapidly since the late 1960's, but of these only palm oil is at all labour-intensive. Employment creation is a problem common to other countries discussed here, and Indonesia's population growth rate is somewhat lower than that of the other tin producers, as the result of a successful birth control programme. It has the added dimension, however, of involving considerable resettlement of population, and under the latest five year plan (Repelita III, 1979-83) the transmigration programme aims to move half a million families to the outer islands.

The present plan and its predecessor have emphasised income distributional targets, stressing food production and employment creation, whereas Repelita I (1969-73) was concerned primarily with reducing inflation and stabilising the balance of payments. Foreign aid under the Inter Governmental Group for Indonesia was an important component of the military government's programme. External financing covered half of the development expenditure under Repelita I, and one-third under Repelita II. Foreign investment was encouraged too, under the Foreign Investment Law of 1967, the Dutch having been expelled by Sukarno in the late 1950's from their main economic activities (including rubber estates and the tin mines). Palmer (1978) argues that this period of 'stabilisation' has led to the creation of an 'exogenous', capital-intensive economy which has grown at the expense of the high-employment, high-linkage 'endogenous' economy (which consists of subsistence agriculture, traditional agricultural exports, tin mining and domestically owned services and manufacturing) and which has done little to alleviate poverty. The 'green revolution' technologies in rice have increased landlessness and redistributed income regressively, while subsidised food imports under the aid programme have lowered the income of local farmers by depressing the rice price. Local small scale manufacturing, she argues, has been harmed by tight-money stabilisation policies at the same time as foreign investment has been encouraged, and imports have expanded with the high rupiah and with the supplies under the aid programme.

Thus severe problems remain for Repelita III if growth is to be more broadly based. The Indonesian rupiah was devalued by a third in 1978 in order to encourage non-oil exports and to reduce import penetration. Manufactured exports are still minimal compared to those of other Asian countries, and the share of manufacturing in GDP is very small. One aim of the plan is to encourage labour-intensive manufacturing.

Most minerals are mined outside of the main areas of population, copper and nickel being the most important ones after tin. Almost all tin is obtained from the islands of Bangka, Belitung and Singkep, off the coast of Sumatra. Chinese immigration, as in Malaysia and Thailand, was associated with the tin industry. The Chinese today number about four million (only about 3 per cent of the population), but they are important in business and commerce. Many were massacred as part of the military takeover in 1966 when they were associated with an unsuccessful communist attempt to seize power. Like Malaysia, Indonesia has a declared aim of restructuring its economy in favour of indigenous Indonesians ('pribumis'). At present 20 per cent of equity in new joint ventures must be in pribumi hands, though it is hoped eventually to raise this to 51 per cent. Like the Malaysian bumiputras most individuals in the pribumi community in Indonesia lack capital, and a state holding company, Dana Reksa, buys equity on their behalf, as does Pernas (the National Corporation) in Malaysia. Apparently these targets are administered rather flexibly, but foreign investment is still substantially less than in the mid 1970's. (<u>Financial Times</u>, 10.10.79).

## 3.4 THE MINOR EXPORTERS

Of the ten largest producers of tin listed earlier in Table 3.1, the UK and South Africa consume all of their own output of metal, though South Africa ships some concentrates abroad for smelting. Of the four remaining minor producers listed, Australia and Brazil consume a substantial part of their metal output too, and indeed Brazil is unusual among developing countries in having a major tin plate manufacturing industry. Some economic characteristics of Brazil, Australia, Nigeria and Zaire are set out in Table 3.8. Brazil has *per capita* income 50 per cent higher than Malaysia's and three times that of Thailand and Bolivia. With a large domestic market the economy is the least open of all the tin exporters and minerals are relatively

TABLE 3.8
ECONOMIC CHARACTERISTICS OF THE MINOR TIN EXPORTERS, 1978

| | Australia | Brazil | Zaire | Nigeria |
|---|---|---|---|---|
| Population (millions) | 14.2 | 119.5 | 26.8 | 80.6 |
| Land area (thousand square kilometres) | 7687 | 8512 | 2345 | 924 |
| Gross National Product per head | $7990 | $1570 | $210 | $560 |
| Average Annual Growth GNP per head 1960/78 | 2.9% | 4.9% | 1.1% | 3.6% |
| Average Annual Population Growth Rate 1970/78 | 1.6% | 2.8% | 2.7% | 2.5% |
| Average Annual Inflation Rate 1970/78 | 12.8% | 30.3% | 26.2% | 18.2% |
| Shares in GDP of - Agriculture | 5% | 11% | 27% | 34% (1977) |
| - Manufacturing | 19% | 28% | 7% | 9% (1977) |
| - Mining & Quarrying | 4% | 1% | 10% (1975) | 30% |
| Share of Exports in GDP | 16% | 7% | 28% (1977) | 31% (1977) |
| Share in (1977) Exports of | | | | |
| - Manufactures | 21% | 26% | 8% | 1% |
| - Agricultural Products | 45% | 64% | 21% | 6% |
| - Fuels and Minerals | 34% | 10% | 71% | 93% |
| (of which tin) | 0.3% | 0.2% | 0.6% | 0.2% |

Sources and Notes

(1) As for Table 3.7.

unimportant in total exports. Brazil attracts attention as a source of new material discoveries, however, and the Brazilian government attaches much importance to the development of new mineral production (39). As a large newly industrialising country (for example, with the fifteenth largest steel industry in the world) Brazil is a heavy consumer of minerals and in the late 1970's it was importing minerals roughly to five times the value of the minerals exported. The rises in the price of oil in the early 1970's caused severe balance of payments problems and this, too, was a stimulus to government interest in minerals. Brazil is at the moment the largest exporter of iron ore (mainly to Japan) and the third largest of manganese, as well as having one of the world's biggest reserve of bauxite. Oil is produced too, but meets less than half of domestic needs. Foreign investment is encouraged though on fairly stiff terms and normally in joint ventures. Companies involved include British Petroleum, Bethlehem Steel (in manganese), Alcan (bauxite), Metallgesellschaft (nickel) and Mitsui and Sumitomo (chrome). In spite of Brazil's large geographical area, 90 per cent of the population live on the narrow coastal strip and most mineral discoveries have been further inland. Tin was found in the remote Rondonia region of the Amazon which has alluvial deposits and contains about two thirds of the country's reserves, and in consequence tin production doubled during the 1970's. Domestic smelting capacity has been installed in excess of current production and concentrates are imported for re-export as metal.

Australia has experienced a boom in mineral discoveries and mineral exports in the last twenty years and minerals have begun to rival the country's traditional food products as the major export. By the mid 1970's Australia had proved to have the world's largest reserves of bauxite, together with major deposits of iron ore, copper, coal, lead and uranium (40). The effect of the minerals boom on other export sectors and on import-competing industries via the appreciated exchange rate has been much discussed in Australia (see chapter two), and Australia's manufacturing sector remains small. It is smaller as a percentage of GDP than Brazil's and not much larger than that of Thailand. Australia's tin is found both in alluvial and primary deposits.

Zaire and Nigeria both depend heavily on minerals for their export earnings. Zaire, formerly the Belgian Congo, has a long history as a copper exporter and tin is of minor importance. Nigeria's exports are mainly of petroleum and, again, tin is a minor product. Petroleum exporting is a fairly recent activity for Nigeria, however, and in 1960 the relative shares of agriculture and minerals in total exports were reversed.

Zaire's turbulent history following independence in 1960 has been closely linked to copper and to the country's largest mining company, the Union Minière. Controlled by Belgian interests, though with a large stake held by the colonial government of the Congo (and then by the newly independent government), Union Minière was a supporter of the short lived secession of the southern copper province, Katanga. Providing about half of the government's revenue and 70 per cent of the country's export earnings, Union Minière was nationalised in 1967. Protracted negotiations and legal action led first to the Belgian owners being granted generous management contracts (which included a large element for compensation) and finally (in 1974) to full compensation. Since then an active policy of encouraging foreign investment has been followed and new entrants have included Charter Consolidated (the foreign partner in the federal Malaysia Mining Corporation) and Lonrho. Lanning and Mueller (1979, p.256)(41) claim that the government's preoccupation with obtaining a share for its supporters of the surplus from mining has led to a total neglect of the rural sector, where live nearly 80 per cent of the population, and whose real income has been declining while food imports increase. Transportation of copper has been southwards rather than through the main centres of population and Zaire suffered greatly from the closure of the railway through Angola during the Angolan civil war. In a study of the economic impact of the copper industry on Zaire's economy, Raymond Mikesell (in Pearson and Cownie, 1974, chapter eight) found that in 1970 the state company, Gecamines (Union Minière's successor which also produces some cobalt, zinc and silver) spent 79 per cent of its total receipts locally. This high figure was made up, however, of 59 percentage points as payments to the government and local linkage effects were slight. Mikesell noted that the company prided itself on adopting the most advanced capital-intensive techniques in spite of the abundance of labour in the country. A further effect of the relatively high wages in mining was that mine workers' consumption expenditure has a high import content. The government has used its large copper revenue mainly for large scale public works, with little expenditure on transport.

The Nigerian oil industry also pays out a large proportion of its export earnings in domestic expenditure. In 1971 it was estimated that this was approximately 70 per cent (42) and the net social gain (i.e. the gain taking into account the social opportunity cost of the resources used) was nearly 60 per cent of gross output (43). The gain is predominantly in the form of payments to the government, as a result of taxation and equity participation, and is of quite recent origin. In 1963 (in the early years of the industry) net social gain was only 27 per cent. Oil revenue provides about 75 per cent of the government's income and heavy emphasis has been placed on agricultural development, agricultural exports having suffered severely from the increase in petroleum exports (as in many other mineral economies). Backward linkages are not negligible (they are some 10 to 35 per cent of total expenditure) but they mainly represent purchases from foreign owned firms. Forward linkages are mainly prospective, and final demand linkages are small, given the tiny (1 per cent) share of local wages and salaries in the industry's gross output value. Foreign oil companies normally work in partnership with the Nigerian National Oil Corporation.

In Nigeria tin is found in the Plateau area in the centre of the country, whereas much of the oil has been found in the coastal areas and offshore. Tin in Zaire is found in the south and south east. In both countries the deposits are alluvial.

NOTES

(1) Though the USSR did disrupt the world market by sales during the 1957-61 period.
(2) Strictly speaking his figures were for the Federated Malay States only, but that includes the two states which produce 90 per cent of Malaysia's present output.
(3) Chhabra, Grilli and Polak (1978, p.23) estimate short run demand elasticities for various groups of consumers ranging from -0.1 to -0.5. They had little evidence on longer run elasticities, estimating a range of -.0.25 for developing countries to -1.60 for developed countries other than the US, Western Europe and Japan (for whom almost no estimates were given, however).
(4) This figure of the size of the surplus is from ITC, MSB, February 1980.
(5) These details about the US stockpile are from Fox (1974) chapters 11 and 15.

(6) The commercial sales in 1973-74 followed a previous suspension of commercial sales since 1968.
(7) TI December 1978. This discussion of consumption also relies heavily on articles in TI, January 1979 and 1980, September 1978 and the TI article by Barry (1978). The paper 'Tin and Its Competitors' by W. E. Hoare in ITC (1974) is also useful.
(8) Such a study is being conducted at present (1980) by the ITC to complement its recently completed study on production (Engel and Allen, 1979). The production study is discussed further in chapter six.
(9) On this criterion thirty-seven LDC's had export earnings which were more unstable. The indices for Malaysia and Indonesia were in the 4-5 per cent range, also for the 1959-71 period, roughly the same range as that of the UK. Although Hallwood constructed price instability indices for nineteen commodities, unfortunately they do not include tin.
(10) See Thoburn (1977a) and chapter six (section 4).
(11) The standard work is Fox (1974). See also Khan (1979) for a description of its operation.
(12) See also Bosson and Varon (1977, pp.109-10). Indonesia is making increasing use of long term contracts which somewhat reduce the size of the 'free' market.
(13) The revenues were discounted back to 1956 using a 5 per cent real discount rate.
(14) Essentially this loss occurs because price control in a demand shift market, assuming the supply function has a position slope, reduces price when supply is high and raises it only when supply is low.
(15) The fifth agreement provides that producers will make compulsory contributions totalling 20,000 tonnes of metal or its cash equivalent. Theoretically another 20,000 tonnes could be contributed by consumers on a voluntary basis, though only the USA has large stocks of metal which it could contribute.
(16) This is in the assumed absence of export control, however. For the 1960-63 period it would have had to have had 100,000 tons.
(17) The price range has been denominated as Malaysian ringgit ($) per picul (60.48g) since 1972. Previously it was set in £ per ton, but the floating of the sterling exchange rate meant that some changes in the £ price of tin were irrelevant to other consumers and to producers.
(18) For example, by Harris (1978, p.10).
(19) The classic description of tin mining technology is Mantell (1970) chapter five.
(20) The largest gravel pump mine is Gopeng in Malaysia. This treats about 800,000 cubic yards a month, a throughput larger than that of most dredges.
(21) In principle open cast methods also could be used to mine tin from hard-rock deposits, as in coal for instance, but in practice it is normally used for alluvial tin.
(22) That is to say that multiple gravel pumps could operate most dredging properties other than those on swampy ground. Gravel pumps have a monopoly of smaller deposits, hillside deposits and of larger deposits where there exists tenacious clay or pinnacled limestone.
(23) For a discussion of other alternatives to offshore dredging see Knights (1980), who analyses methods of digging deeper than 50 metres.
(24) See Thoburn (1977b, chapter eight) for a case study of the growth of the Malaysian engineering industry in its relation to the primary export sector.

(25) In view of the lack of ownership links between tinplate and mining and smelting, possible links between the tinplate and the can manufactures were not investigated.
(26) Lanning and Mueller (1979, chapter fifteen) also give a useful description of the increased scale of mining operations in other metals.
(27) For discussions of optimal dredge size see Cullen and McDonald (1974) (who argue that 60,000 cubic yards per month is an effective maximum) and Donkers and Hauw (1975) who argue that efficient larger dredges are possible.
(28) They could not cope with the heavy vegetation and wood found in many Indonesian deposits. Their usual bucket size was about five cubic feet.
(29) The lack of a close relation between bargaining strength and industrial structure is also suggested by Radetzki (1977) in a study of bauxite, iron ore and copper, but he is more concerned (except for fully vertically integrated operations) with bargaining between producers and consumers than between foreign companies, local firms, and host governments.
(30) Thoburn (1977b) gives information about Malaysian development on which this account relies heavily. More recent information derives from the Mid-Term Review of the Third Malaysia Plan 1976-80.
(31) See Wong (1979) for further details of ASEAN.
(32) The most useful source on Bolivian economic development in the past is Zondag (1966). Geddes' (1972) biography of Patino is also informative on the country as a whole, as is the charming memoir of Bolivia by the former UNDP representative Margaret Anstee (1970). Gillis's (1978) excellent study of the Bolivian mining industry unfortunately has little on the general development of other sectors. Lloyds Bank (1980) is helpful on recent events.
(33) See the National Union of Mineworkers (1979) report on Bolivia, pp. 14-15.
(34) In 1979 the growth target for GDP was 7.7 per cent per annum as against an estimated actual increase of 2.1 per cent (Lloyds Bank, 1980 p.12).
(35) Formed in 1969, the Andean Pact includes Colombia, Ecuador, Peru and Venezuela and aims to promote regional industrialisation and free trade.
(36) This account relies heavily on the World Bank (1980a) country study on Thailand, which is based on the findings of a mission in 1977. Country studies by the Bank on the other tin producers are not available to the public. Ingram (1971) is also useful. For a rather different perspective on Thailand's political and economic development see Elliott (1978).
(37) Mining is barely mentioned in the World Bank report on Thailand, for example.
(38) The most useful general source on Indonesia is Palmer (1978), though she is very critical of many of the country's recent developments. Sumitro (1978) is informative on future perspectives, and the latest five year plan (Repelita III) is described in the Financial Times survey of Indonesia (10.10.79) and in FEER (2.2.79). This account draws heavily on these sources.
(39) See Lloyd and Wheeler (1977) from which the information here on Brazil is taken.
(40) See McKern (1976, p.5). McKern gives a detailed discussion of the Australian minerals boom.

(41) This section on Zaire is derived mainly from chapter ten of Lanning and Mueller.
(42) This section draws on R. K. Meyer and S. R. Pearson's study of the contribution of petroleum to Nigerian economic development in Pearson and Cownie (1974, chapter seven).
(43) The notion of net social gain was discussed in chapter two (section 3).

# 4 The Development of Multinationals' Involvement in the Industry

This chapter traces the growth of MNC participation in tin from the nineteenth century to the present, as a prelude to the discussion in chapter five of recent policy changes in the producing countries. Section 1 looks at the early history of mining in Asia and Africa, and section 2 considers in more detail how tin dredging became a vehicle for foreign involvement. Section 3 outlines developments in Bolivia and section 4 the initial attempts at financial integration in the industry. Section 5 analyses changes in MNC involvement since the Second World War, particularly the decline of specialist foreign tin companies and the entry of major mining multinationals.

## 4.1 THE ENTRY OF FOREIGN FIRMS IN ASIA AND AFRICA

In none of the three main producers in Asia did tin mining depend for its early growth on foreign firms. In Malaysia firms initially faced strong competition from well established mines run by immigrant Chinese. This was also true in Thailand to some extent. In Indonesia much of the industry was developed by the State, though again there was a prior history of Chinese involvement in mining.

In southern Thailand immigration by Chinese in response to mining opportunities had been taking place for several hundred years prior to the industry's expansion at the turn of the twentieth century, and Thais themselves rarely mined. Indeed, Chinese from the Malay peninsula were actually invited to invest in tin mining by southern provincial governors in the late nineteenth century in order to raise tax revenue (Wit, 1971, p.9). In 1892 the system of mining licences and export royalties was revised, partly to encourage Western companies (1), but there is little published information about the activities of Western companies before offshore dredging was introduced in 1907. In Malaysia, Wong's (1965) economic history of the industry to 1914 provides evidence of the interaction of local and foreign firms. By 1870 the great Larut tinfield in Perak had been opened up by Chinese, who had long since pushed out of business the indigenous Malay miners. British control was extended into the Malay Peninsula in 1874 on the pretext of preventing clan warfare in the Chinese mining community, and a massive expansion of tin exports began. Malaya soon overtook Britain, the then Dutch East Indies and Australia to become the largest world producer by 1879. By 1896 Malaya produced about 60 per cent of world output. In the expansion to the mid 1890's Western firms played little part. There were many company flotations, but there was also a high failure rate as a result of heavy overhead costs, labour problems and a management unfamiliar with alluvial mining. One of the few successful companies was one operating an underground deposit. Chinese mines were open pits and used very large labour forces, roughly ten times more workers being employed per cubic yard of throughput than in a modern gravel pump mine (Thoburn, 1977b, p.127). Most Chinese workers worked on a profit sharing basis - on 'tribute' - which was a method disliked by British firms, and Chinese employers were able to withhold labour from foreign companies through their involvement with secret societies.

Government encouragement, especially the suppression of the Chinese secret societies in the 1890's, helped western companies but their rise owes more to the difficulties faced by Chinese mines in coping with deeper deposits and lower grades, and with an increasingly severe labour shortage. The latter was partly the result of the tin market recession from 1889 to about 1899, which reduced immigration, and also factors reducing emigration from China such as the good harvest in 1897 (Wong, 1965, p.172).

The search for labour saving techniques became important for the whole industry. It was intensified from 1896 to 1907 by the authorities' closure of the land application books in the largest tinfield, the Kinta Valley in Perak (Wong, 1965, pp.114, 220), which diminished opportunities of working new land with traditional methods. Also the new rubber industry competed for labour from the first decade of the twentieth century, though much of its demand was met by importing Indian workers.

Western firms were now at a considerable advantage since the new methods eventually to be adopted - gravel pumping and dredging - were techniques used to mine other minerals elsewhere. Thus gravel pumping was used in Cornwall, the centre of tin mining in England, for mining china clay (Allen and Donnithorne, 1954, p.157) and precious metals were mined in Australia by dredging. What is interesting is that the first major new technique - gravel pumping - which is of moderate rather than low labour intensity (2) and which helped to establish promised present day British companies such as Osborne and Chappel in the industry - should have been taken up by the local miners and has formed the basis of the local sector in Malaysia since about the 1920's to the present. Chinese firms in fact adapted the gravel pump method from the suction dredges which western companies had experimented with at an earlier date.

In Thailand, the date of introduction of gravel pumping is not known (Pajon Sinlapajan, 1969). It does not seem to be known either whether gravel pumping was introduced to Thailand by foreign companies. More likely it was introduced by Chinese from Malaya, who still maintain various contacts with the Thai mining community. In any case the degree of competition between foreign and local mining firms before the First World War would have been less intense in Thailand than in Malaya since many of the first foreign dredging companies were operating offshore and therefore not competing for leases.

In colonial Indonesia, the Dutch East Indies, almost all output came, as today, from the islands of Bangka, Belitung (then 'Billiton') and Singkep, with nearly two-thirds from Bangka, one-third from Billiton and only minor production from Singkep. The mines in Bangka had been operated directly by the Dutch colonial government through its Bureau of Mines since the takeover from the British in 1816 by using immigrant Chinese labour. Billiton island, on which tin was first found by Europeans in 1851, was worked by the Dutch Billiton company from 1860, first as the sole concessionaire and then (from 1923) in 37.5/62.5 per cent minority partnership with the colonial government (ter Braake, 1944, pp.37 and 46). Another Dutch company worked Singkep (Jones,

1925, p.23) but its concession was taken over by Billiton in 1934. Most of Bangka's ores were smelted on the island, while those from Billiton and Singkep were shipped to Holland for smelting by a subsidiary company of Billiton, which also smelted part of the Bolivian production (ter Braake, 1944, p.45).

Unlike the British mining companies in Malaysia, the Dutch seem to have had little difficulty in using Chinese labour, even for the highly labour-intensive open cast methods. Indeed, in this regard they were continuing a tradition started by the local rulers in recruiting Chinese (ter Braake, 1944, p.36). On Billiton island the company predominantly had workers working under subcontractors using a 'tribute' system, and, workers sold ore to the company with certain guarantees about minimum payment. Recruitment was organised for a while by the company itself but by the end of the nineteenth century it was left to relatives of existing workers (3). On Bangka there had been in fact a tradition of tin mining by immigrant Chinese which long antedated the Dutch (4). In the late eighteenth century there were two major centres of tin mining - Phuket in Thailand and Bangka island. After a discovery of tin on Bangka in 1710, rapid expansion took place in output and in Chinese immigration, the period 1750-1780 being the peak of production. In the 1760's output was thought to have been about 3500 tons, roughly the same as that of Cornwall at the time, and mostly exported to China. Techniques were mainly large open cast mining similar to those of nineteenth century Malaya and the Malayan Chinese organisation of the labour force in kongsis and of financing by advances were also antedated by Bangka. When the island was taken over by the Dutch, many decades passed without significant technical change; and the widespread use of mechanical power was getting underway only by the end of the nineteenth century.

In Burma, then part of the British Empire, Jones (1925) records a number of European companies in operation, and gravel pumping being quite widely used. At Tavoy, in the south of the country, another suction dredge operation was pioneered by a Sterling company. Similarly in Indochina some European involvement was recorded. In both countries the First World War had provided a stimulus to mining because of the demand for tungsten for armaments, tin and wolframite (tungsten) being found together in many areas in the two countries.

In China, which in the 1920's was the fifth largest producer, mining remained under local control and attempts by European firms, even to buy concentrates for smelting, were firmly resisted. Over 90 per cent of output at that time came from the Kochiu district in Yunnan province and the methods as described by Jones (1925, pp.240-245) sound similar to the older open cast methods used in Malaya in the 1870's.

In Africa, Nigeria by the 1920's was accounting for over 80 per cent of the continent's output, having reached the 1000 ton level only in 1911, with no significant output before 1907. Although local miners had previously worked at a very small scale, the growth of the industry was predominantly associated with British companies, using a variety of hydraulic techniques. Apart from minor output from the Belgian Congo, South Africa was the other important African producer.

In 1920, over 60 per cent of world tin ore output was being smelted in the non-industrialised producing countries - 41 per cent in Malaya, 11 per cent in the Dutch East Indies, and 9 per cent in China. Australia (3 per cent) also smelted part of its own output, while most of the rest of the world smelting was carried out in the UK (20 per cent) and the USA(13 per cent). Only minimal amounts of domestic mine production in Bolivia and Thailand were locally smelted (Jones, 1925, p.25). In Malaya the smelting was divided roughly equally between the Straits Trading Company, founded by Europeans as an independent venture in the late nineteenth century, and Eastern Smelting, which was set up by local Chinese. By 1911 Eastern had been taken over by British interests (Allen and Donnithorne, 1954, p.160).

Rather ironically, Straits Trading at a later date was taken over in turn by Chinese financial interests. Smelting in Malaya was helped by a prohibitive duty on the export of tin ore from 1903 from which exports to the UK were exempted (from 1904) (5).

## 4.2 THE DEVELOPMENT OF DREDGING

The basis for really rapid growth in foreign investment in Thailand and Malaya, and for the subsequent strength of foreign companies, was dredging. Dredging's high capital costs for many years effectively prevented local entry, especially since limited liability companies among Thai and Malaysian miners did not develop widely until after the Second World War.

The early dredging companies in Malaya and Thailand frequently came either from Cornwall or Australia, whose own tin industries had declined partly through reserve exhaustion and partly through falling prices during the late nineteenth century depression already mentioned, as world output growth outstripped even the rapid rise in demand from tinplating in Britain and the USA. In Thailand the world's first tin (bucket) dredge was put into operation in 1907 by an Australian company, Tongkah Harbour, off Phuket Island on the south-west coast, and by 1910 five dredges were in operation (Pajon Sinlapajan, 1969; and Rachan, Pow and Champion, 1969). The Siamese Tin Syndicate (still, like Tongkah Harbour, an important company today) was floated in 1906 to mine in the same area (Fairmont State Ltd., AR, 1978) and is said by Wong (1965, p.210) also to have been originally Australian. While the Australian companies appear to have 'migrated' to Thailand, British companies maintained links with Cornish mining (Allen and Donnithorne, 1954, p. 156). Also, there were close links from the beginning between Malaya and Thailand. Thus Renong Dredge Company from Australia, formed to mine inland in Thailand in 1906, owned properties in Malaya to which it transferred dredges in the 1920's as the Thai deposits reached exhaustion (Jones, 1925, p.217). Kamunting Tin, one of the first dredging companies in Malaya, operated also in Thailand until the early 1970's. By 1928, the first year for which sectoral output statistics are available for Malaya, dredging was well established, producing about 30 per cent of total tin output in both countries. By the Second World War the shares were nearly 50 per cent in Malaya and over 60 per cent in Thailand (6).

In Indonesia too, techniques changed as lower grades of ore were encountered after the initially rich deposits were worked out. The process of technological change is reasonably well documented, at least for mining on Billiton island (by Molloma, 1922). By the early 1880's fears were being expressed about the exhaustion of Billiton's reserves and an intensified exploration programme was started. Significant changes in techniques date from the end of the nineteenth century, and by 1916 mechanical methods were producing a larger output than manual ones. The suction dredge was introduced from Australia and the Straits in 1906 and, as in Malaya, the floating pontoon seems quickly to have been dispensed with and the technique was developed as gravel pumping. The first bucket dredge on Billiton dated from 1920, though Zaalberg (1970) records that a sea dredge was floated off Singkep in 1911. If the much greater richness of deposits on Bangka relative to Belitung island in the present day is any guide, the need for technical change on Bangka would have been less pressing. Nevertheless it seems that the techniques spread quickly and by 1939 dredges produced 40 per cent of the country's output (Fox, 1974, p.32). In Indonesia though, there was no interplay between choice of technique and ownership and control, since the Dutch colonial government and the Billiton company had been given an effective monopoly of tin mining; so gravel pumps were used primarily to mine small deposits.

In other alluvial tin producers dredging seems to have spread rapidly, and Jones' survey of the world tin industry published in 1925 records dredges in operation in Burma, Nigeria and Australia. In Nigeria and Australia, however, the interesting interplay between dredging and the growth of foreign companies was lacking, as in Indonesia. In Australia the industry was developed mainly by Australian companies and in Nigeria gravel pumping was used predominantly by the British firms. In the Congo, even after the industry was developed rapidly with Belgian capital in the 1930's (Fox, 1974, pp.52-53) dredges were unimportant, the deposits either being very small or very near the surface and amenable to simple open cast methods.

## 4.3 THE EARLY DEVELOPMENT OF THE INDUSTRY IN BOLIVIA

The interest of the Bolivian case in this chapter is much less the role of foreign companies than the fact that Bolivia was the starting point of the development of a particularly influential multinational, the organisation founded by the Bolivian Simon Patiño and based in the United States after 1924.

Along with Malaya, Bolivia was one of the earliest sources of supply for the expansion of tin consumption at the end of the nineteenth century, and from 1905 to 1920 she was the second largest producer. Tin production in Bolivia grew rapidly in the 1890's, coinciding with the exhaustion of the main silver deposits and a drop in the market for silver, hitherto the largest export. Tin had always been mined as a byproduct of silver and the well established silver mining cities Oruro and Potosi on the high plateau formed a nucleus for the growth of tin. Growth was aided in the 1890's too by Oruro being linked by rail to the Chilean coast. Much capital for the expansion came from silver interests (7) but some foreign capital also entered the industry, especially from Chile. In Jones' (1925) account of the industry in the 1920's many Chilean and some British and French companies were listed.

Patiño in fact was the instrument whereby this foreign control was greatly reduced. By the 1930's Patiño controlled roughly half of Bolivian tin production, another 10 per cent being controlled by the Hochschild company and 15 per cent or more by Aramayo (8). Aramayo has been described as 'Swiss, with Bolivian capital' (Jones, 1925, p.327) and Fox (1974, p.63) refers to Hochschild as an individual whose capital connections were 'unclear'. These three groups continued to dominate the industry until their nationalisation in 1952.

Patiño's growth originated in his finding an exceptionally rich deposit in his first mine, which by 1910 was producing some 10 per cent of world production. A series of acquisitions of French, British and Chilean owned companies, and railroad and banking interests, led to the formation of a group - Patiño Mines and Enterprises Incorporated - incorporated in the USA. The decision to locate in the USA is ascribed by Klein (1965) to a wish to tap US sources of capital, and Patiño had indeed used credit facilities skilfully to build up his group (9). It also greatly lessened, of course, the Bolivian government's control over the company's activities, and the reorganisation accelerated a process of acquisition which was to lead to interests in Asia, Europe and North America.

Besides the Patiño interests in Bolivian tin, banking and railways, one very important diversification had already taken place by 1924 - the group had acquired control of Europe's largest tin smelter (Williams, Harvey of Liverpool) and in partnership with the National Lead Company of the US, at the time the world's second largest consumer of tin.

The National Lead-PMECI combination represents a grouping of consuming and mining interests unusual in the tin industry, even today. Their threat in 1916 to build their own smelter had allowed them to buy half of Williams Harvey's equity, and they acquired the rest soon after; the first half being acquired 2/3:1/3 and the second half each. National Lead was also an active participant in the takeover of the largest mine in Bolivia, previously owned by Chileans. Patino had also acquired an interest in a German tin smelter, to which some of his ore had been sent before the First World War.

## 4.4 INTEGRATION IN THE TIN INDUSTRY TO THE SECOND WORLD WAR

During the late 1920's and the 1930's a process of consolidation occurred, bringing about a considerable degree of horizontal integration and strengthening the links between mines and smelters. The two main developments were the formation in 1925 of what was to become the London Tin Corporation, and the establishment of Consolidated Tin Smelters in 1929.

By 1937 the LTC controlled through its subsidiary Anglo-Oriental twenty companies in Malaya producing about a third of the country's output and roughly half that of the foreign sector (Allen and Donnithorne, 1954, p.157). Companies in Thailand came under LTC control too, such as Tongkah Harbour, and also in Burma, while the LTC controlled Associated Tin Mines of Nigeria which accounted for half the country's output. The LTC did not normally have majority ownership but control could be exercised since other shareholdings were dispersed, besides which Anglo-Oriental acted as managers. Control was operated through a combination of management and interlocking directorships, a pattern that persisted into the 1970's (10).

Consolidated Tin Smelters was formed by Patiño, following his group's acquiring control of the Williams Harvey smelter in England (11) which then smelted much of Bolivian production. Besides two other smaller British smelting companies, control of Eastern Smelting in Malaya was secured (12) and CTS also held 'substantial' interests in the Dutch smelter at Arnhem which by the early 1930's was smelting most Indonesian output (which previously had been smelted in Malaya) (Allen and Donnithorne, 1954, p.160). Of the major smelters there thus remained outside CTS only the Straits Trading Company in Malaya.

Between these two groups there were a number of cross holdings, though information on their quantitative extent is not readily available. Yip (1969, p.184) mentions that Anglo-Oriental held a substantial holding in various Malayan dredging companies, particularly as a result of acquisitions made after share prices fell following the 1929 crash. Geddes (1972, p.228) cites British press reports to the effect that Patiño owned 80 per cent of the shares in certain large LTC operating companies, though only 10 per cent of holdings (held via CTS) have persisted into recent times.

The LTC-CTS combination was a major influence for the establishment of tin production control schemes in the 1930's and the evidence for cooperation between the two groups is generally accepted (13). It also appears to have worked against the interests of Malaya, whose quota was underassessed (especially relative to Thailand), but the LTC did not object because of its interests in higher cost production elsewhere (Allen and Donnithorne, 1954, p.162).

Nevertheless, in spite of this integration, the market position of producers outside the group (the majority in Malaya and Thailand and half in Nigeria and Bolivia) was not significantly weakened. Other Bolivian producers in the 1930's undoubtedly gained from restriction in a situation where Bolivia might have gone out of production altogether. The prospects of the Malaysian mines outside Anglo-Oriental's control were damaged, but no more than those of Anglo-Oriental's own mines. More important, the existence of the Straits Trading Company as a large independent smelter, though with some holdings in mines in the Osborne and Chappel group, meant that independent miners were not subject to significant monopolistic pressure and most marketing remained independent of the large groups. Patiño's acquisition of smelters was clearly an attempt to secure his market position, and virtually no integration forwards into tin using industries occurred, nor did tin users themselves acquire interests in mining or smelting other than National Lead's cooperation with Patiño prior to 1929.

## 4.5 CHANGES IN PARTICIPATION BY MNC'S SINCE THE SECOND WORLD WAR

In the 1950's approximately three-quarters of Bolivian tin production was nationalised and complete nationalisation was carried out in Indonesia. In the 1970's the governments of Malaysia and Thailand took measures to exert greater control over the industry, as did Nigeria.

Zaire too has taken a stake in its tin companies though its main concern is copper. In the USSR and China, of course, production is under state control though neither country had any previous experience of foreign companies in their tin industries. In the main western industrial countries producing tin - Australia, South Africa and the UK - the industry remains in private hands but with various multinational links.

After the Second World War and the granting of independence to Indonesia, mining on Bangka and Belitung was put under the control of the Billiton company for five years from 1942, but on the expiry of the agreement in 1953 the Indonesian government assumed the running of tin mining on Bangka; and Billiton was excluded from Belitung and Singkep in 1958 (Fox, 1974, p.35). Since 1968 mining has been controlled by the State enterprise P. N. Timah (now P. T. Timah). The benefits of this nationalisation (and that of Bolivia) will be taken up in the next chapter, which will also consider the effects of the policy started in 1967 to re-attract foreign investment under which Billiton has re-entered the Indonesian tin industry together with two new companies.

In Malaysia and Thailand various changes occurred which affected the relative bargaining positions of the companies and the host-country prior to the major policy changes in the mid-1970's which are discussed in the next chapter. First, several new multinationals have entered the industry. In Malaysia Conzinc Rio Tinto Malaysia, an Australian owned subsidiary of the (British) Rio Tinto Zinc Corporation, started with a single dredging operation in the early 1970's, and has embarked subsequently on several joint ventures with individual state government mining organisations. More important, Charter Consolidated bought into the Tronoh group in 1965 and in 1975 controlled companies producing nearly 10 per cent of total Malaysian tin output; thus rivalling the LTC, whose companies accounted for 15 to 17 per cent (14). CC, through Tronoh, also controlled Aokam Tin in Thailand, though LTC also had influence in Aokam through Malayan Tin's 23 per cent holding (Malayan Tin Dredging (M) Bhd, AR, 1978). Charter is now the minority partner in the federal Malaysia Mining Corporation.

In Thailand the Dutch company Billiton, from colonial Indonesia, entered the industry in partnership originally with the American company Union Carbide. UC had been especially interested apparently in certain byproducts of tin mining and tin smelting and in 1965 had started the construction of the THAISARCO smelter in partnership (UC holding 70 per cent, later 60 per cent) with a Thai company, which had previously secured a large offshore mining concession (Offshore Mining Organisation, 1978, pp.12-13). Billiton was invited to join the venture for its mining expertise in about 1970 on a 50/50 basis, the Thai partner having left (Interview, Billiton Thailand, October 1978), and mining and exploration operations were organised under two companies (TEMCO and EMCO) with substantial offshore leases. UC left in 1974, following difficulties with (and eventually revocation of) the mining leases, giving Billiton 100 per cent ownership of the mining operations and the smelter (which smelts almost all Thai output, which had previously been sent to Malaysia).

The entry of RTZ (CRM) and Charter, and Billiton too to some extent, represents a significant change in industry structure and in the foreign bargaining position. In the early twentieth century it was quite common for mining companies to be specialised on a single country, but the subsequent pattern has been of diversification using retained profits (15). While the LTC resulted from the horizontal integration of tin mining companies in several countries, even by the early 1970's it had not diversified significantly beyond tin. RTZ in contrast, the largest mining company in Britain, has interests in almost every metal in a wide variety of LDC's and industrial countries and substantial interests in smelting in various non-ferrous metals. Its interest in tin is fairly new, however. CRM has started an attempt to diversify in Malaysia beyond tin mining, and has begun a joint operation with the state's Economic Development Corporation to explore for base metals in Pahang. RTZ's stake in Malaysia through CRM is still small, and it is Charter's which is of greater interest. Charter Consolidated was formed in 1965 out of several disparate companies in the Anglo American Corporation group, the largest South African mining house and indeed the largest mining company in Africa (16). Anglo held in 1979 a controlling 35.6 per cent of the equity of Charter (Charter Consolidated, AR, 1979), which has been one of Anglo's main vehicles for diversification beyond South Africa. Charter has interests, like RTZ, in a wide range of minerals, including lead, zinc, wolfram, potash and gold, and is reported to be trying hard also to develop manufacturing interests (The Times, London, 13.2.80). Its activities in South East Asia generated only 9 per cent of its 1979 revenue.

Billiton too is a much more diversified company than the LTC. During the 1930's it spread from Indonesia tin into investment in bauxite, and after its exclusion from Indonesia in 1958 it moved into lead and zinc. In 1970 it was taken over by Royal Dutch Shell as part of that company's diversification into minerals, a move parallel by other oil companies (UNIDO, 1979, p.203).

Billiton's ownership of the Thaisarco smelter in Thailand supports a major marketing operation by Billiton which handles not only almost all of Thailand's production but also about 18,000 tonnes from Malaysia, Bolivia and China - all in all about 20 per cent of world tin trade (Interview, Billiton, The Hague, May 1978). Among other entrants to the industry it is worth noting that Billiton's temporary partner, Union Carbide, which is the second largest chemical company in the USA, owns a variety of mining interests in the Third World, as does Bethlehem Steel from the US which was trying in the late 1970's to establish an inland and an offshore tin dredging venture in Thailand, presumably to try to secure a tied source of tin (17). Lonrho in 1980 acquired a substantial (17 per cent) stake in Malaysiam Tin in Malaysia (TI, April 1980).

In contrast to these new and diversified multinationals the LTC in 1974 still had 85 per cent of its investment in tin, of which Malaysia accounted for 75 per cent points (18). Other metals and oil accounted for another 10 per cent (Zorn and Leigh-Hunt, 1974).

The second type of change in Thailand and Malaysia concerns the interaction between foreign companies and the local sectors. In both countries the range of techniques in use remains similar to that between the wars, apart from the recent growth of small boat mining in Thailand.

In both countries the industry remains fragmented - in Malaysia small scale gravel pump operations comprise about 90 per cent of the total 800 mines, and small scale operators in Thailand are over 90 per cent of the total (19). In both countries dredging, the main area of foreign involvement (20), has declined in relative importance, falling in Malaysia from about 55 per cent at the end of the 1950's to just over 30 per cent in 1975, and in Thailand from 65 per cent in 1950 to 16.6 per cent in 1976. In Malaysia the dredging decline has been matched by a rise in gravel pumping's share, and in Thailand by gravel pumping and by the emergence of small boat mining (about which more in chapters five and six) from the mid 1970's. The relative profitabilities of dredging and gravel pumping during the post-war period would be difficult to establish without substantial historical research. What can be established is that gravel pumping's profitability is more price sensitive than that of dredging, and their relative rates of return on investment switch within realistic price ranges (21). However, there does not seem to have been a clearly defined upward trend in the real tin price from the mid 1950's until the early 1970's (apart from a brief upward spurt from 1964 to 1967) (22) which would explain gravel pumping's relative expansion; and one would expect rising real wages in Malaysia and Thailand to favour the capital-intensive technique. To some extent it may be an indication that new deposits are small and therefore unsuitable for dredging, or simply that there has been little exploration for new deposits. The latter factor has been greatly stressed in recent Malaysian discussion (at the 1977 National Seminar on the Mining Industry in Kuala Lumpur, for example) and attributed to the adverse effect on investment incentives of 'penal' rates of taxation (Thoburn, 1978b). As a result, much gravel pumping is on land previously worked by dredges (23). In fact, much new dredging investment has taken place in Malaysia since the Second World War and (until the most recent output decline which lasted from 1972 to 1978) it has been in the context of a rising output trend (24). In Thailand much less dredging investment took place after the war and Thailand has missed out almost entirely on the really large capacity 'second generation' of dredges built in Malaysia (25). Another quite striking change in Thailand has been the entry of local firms into dredging using secondhand equipment. Of the 17 dredges in operation in 1976, 12 were working inland, and of the latter nine were in the control of local firms (probably wholly Thai owned). Some estimates based on company reports suggest that in 1977 only about 40 per cent of inland dredging output was produced by companies not wholly local, though until 1978 the entire offshore dredging output was produced by foreign companies (26).

In Thailand, besides the Billiton-Union Carbide operation, there remained by the early 1970's only two large foreign groups, both of long standing. The 'TASK' group, comprising Tongkah Harbour, Aokam, Southern Kinta and Kamunting was under the control of LTC, on whose part Anglo-Oriental acted as general managers (27), and Southern Kinta and Kamunting also operated in Malaysia (28). The other group is the Siamese Tin Syndicate and Bangrin who are grouped under the Fairmont State company in Thailand, the new name of the Faber Merlin group (29), together with the Sichon underground mine (opened 1973) (30). Siamese Tin (and thence Faber Merlin and Fairmont State) retains strong links with the Cornish mining (and property) group Saint Piran, which in 1977 increased its holding in Fairmont to 43 per cent, and there is also a Malaysian branch of Siamese Tin which is 100 per cent owned by St. Piran (31). In the recent past profits from Siamese Tin have been used to finance investment in the Cornish tin mining (32).

In Malaysia direct local entry into dredging has been much less frequent. Indeed, there are only two known cases. In the first, Selangor Dredging, a public company was floated by Malaysian Chinese in the 1960's, and in the other a large gravel pumping company bought second-hand dredges and currently operates two properties in Selangor and Perak. Thus in Malaysia prior to the takeover of LTC by the federal government, to form the Malaysia Mining Corporation, the dredging sector was still quite concentrated. The LTC owned over 25 per cent of the shares of seven dredging companies and over 10 per cent in a further three (Thoburn, 1977b, p.127). Although in no case was majority ownership held, control could effectively be exercised since other ownership was dispersed, and the instruments of control was in providing management services on the pattern established in pre-war years through Anglo-Oriental. The high degree of concentration within dredging in Malaysia is further illustrated by the fact that twelve companies, producing nearly 80 per cent of the country's dredging output had as secretaries or managers either Anglo-Oriental, Osborne and Chappel or Charter Consolidated (or its subsidiary Associated Wines) (Thoburn, 1977b, p.127). In spite of its British origins, the multinationality of Osborne and Chappel is limited in the sense that its activities appear to be mainly in Malaysia. However, it has had some links with LTC and CC through interlocking directorships.

In addition to the post-war changes in control, there have been some changes in ownership. Increases in local purchases of shares in locally incorporated tin dredging companies in Malaysia is well documented over the period 1954-64 (Yip, 1969, pp.348-373) and by 1970 over a third of all dredging company shares in Malaysia were locally owned (Thoburn, 1977b, p.89).

Among the minor producers, by the early 1970's Nigerian tin production was still dominated by foreign companies. Amalgamated Tin Mines of Nigeria, in which the London Tin Corporation had a 30 per cent controlling interest, produced about 45 per cent of the country's output, and identifiably British companies in total produced about two-thirds (33). In Zaire, by the early 1970's according to Fox (1974, p.53) there were only four groups of importance, all working alluvial concessions and all Belgian owned. In 1976 three groups joined to form Sominki, which produces about two-thirds of current output. Géomines, the other group, had become Zairétain, producing about one-third of the country's output, with the Zaire government taking part of the equity (discussed in the next chapter). Géomines' Belgian ownership is apparently independent, whereas Sominki is part of the Cofimines-group. Cofimines is in turn part of the Baron Empain Schneider group diversified into steel, transport and power and employing over 100,000 people (34).

In Australia a large scale revival of the industry in the 1960's as part of a more general minerals boom was associated with the entry of two multinationals, Consolidated Goldfields of the UK and Comin of Canada (Fox, 1974, pp.73-75). CGF, like Anglo American, is of South African origin and tin accounted for 18 per cent of its profits in 1979, according to its annual report. CGF's Ranison underground mine in Australia produces over 5000 tons a year, the largest output of any private tin mine in the world and almost half of Australia's production. CGF also has substantial minority ownership in the Rooiberg underground mine in South Africa (35), the country's largest, and was instrumental in opening the Wheal Jane tin mine in the UK in the 1970's, (though it subsequently closed). Cominco owns the Aberfoyle mine, Australia's second largest. One interesting and recent change in the industry is that the Anglo American Corporation, which owns Charter Consolidated, acquired early in 1980 a 25 per cent interest in CGF (36), though claiming it did not wish to exercise control (The Observer, London, 17.2.80). British Petroleum has also taken a stake in the Anglo group by its bid in mid 1980 for Selection Trust, in which Charter owns 27 per cent (The Observer, London, 13.7.80).

In Brazil, where tin has only been mined commercially since the 1960's a major find in the Rondonian region in the Amazon basin started a rapid expansion. Various foreign companies have been involved including Patino NV (now based in the Netherlands), Billiton, the American firm W. R. Grace (involved in medium scale mining in Bolivia) and NL Industries (formerly the National Lead Company of the USA, which was a partner in Patiño's early ventures). All of Brazil's output is domestically too, which is unusual among Third World tin producers and reflects Brazil's large domestic market size in relation to a still small production by world standards (Fox, 1974, pp.69-71; Engel and Allen, 1979, pp.63-66).

Integration backwards by multinational tin mining companies into mining supplies and mining equipment has been very rare, Anglo-Oriental's former dredge design department in Malaysia being only a partial exception. Forward integration by mining companies into smelting is less rare, but by no means the rule. Billiton's mining interests in Thailand are slight and are somewhat incidental to its major interest in smelting in Thailand and marketing. Its tin interests in Indonesia too are limited to a single mining project. The development of local smelting in Bolivia and Indonesia is part of the increasing involvement of the state in those countries and is dealt with in the next chapter. In Malaysia one smelter, the Straits Trading Company, is now owned by mainly Singaporean interests unconnected with mining (principally the Overseas Chinese Banking Corporation). The most interesting feature of smelting in the early 1970's was Patiño's continuing interest after the nationalisation of its Bolivian mines in 1952. Patino's Consolidated Tin Smelters merged in 1975 with Amalgamated Metal Corporation, a trading firm which the Patiño group had controlled since 1968. As a result of the merger, AMC held controlling interests in the Datuk Keramat smelter (the new name of Eastern Smelting) (50.5 per cent), the Nigerian Smelter (58 per cent) and Associated Tin Smelters in Australia ($33\frac{1}{3}$ per cent). It also continued to control Williams Harvey in England, though this went into liquidation in 1973. AMC was taken over by the German mining multinational Preussag AG of Hanover, in 1978, Preussag obtaining a 76 per cent interest (AMC, AR, 1978). Preussag, though active in many non-ferrous metals, had no previous interests in tin, and this acquisition represents an essentially horizontal diversification. Apart from the former Patiño connection with the LTC (now Malaysia Mining Corporation) tin companies (37), AMC's mining interests are minimal. Its main associated companies are in trading, but it has a 20 per cent interest in a major German copper and lead refiner.

In Malaysia, the other smelter, Straits Trading, has a number of investments in Malaysian tin mines, but these mines are not of great quantitative importance (38).

Over the last decade or so, then, the degree of vertical integration among the MNC's mining tin has generally diminished. Patiño's Amalgamated Metal Corporation which controlled half of Malaysia's smelting capacity besides that of various small producers and the British smelter treating Bolivian ores, has been sold to a company with virtually no tin mining interests. The other Malaysian smelter remains outside the control of the mining industry altogether, and the Indonesian state mining corporation smelts almost the whole of that country's output. Bolivia too is well on the way to State control over the smelting of domestic ore production, as the next chapter will show. Billiton in Thailand is the main exception to this pattern, having a near monopoly of smelting and marketing, though only one mining project. Rio Tinto Zinc wholly owns the Capper Pass smelter in England which now smelts much of the Bolivian output that its domestic smelter cannot handle, but RTZ's interests in tin mining are slight(CIS, 1972). AMC's sale to Preussag has also, of course, reduced vertical integration up to tin marketing stage. Consolidated Goldfields controls one of the largest metal traders, Tennant (active in Bolivia, for instance - see Widyono, 1977, p.36), and from 1979 CGF owned smelting capacity via their Rooiberg company's new smelter in South Africa (TI, October 1979),but, again, CGF are not important in the world tin mining industry.

With regard to integration forward into the finished product there appears to have been little change since Fox (1974, p.14) claimed there were virtually no links between tinplate producers and either smelters or miners. One exception is Billiton, which is a producer of solder. Another is Broken Hill Proprietory which operates large tinplating works in Australia and owns a small underground tin project in Indonesia.

NOTES

(1) This account of Thailand draws heavily on Ingram (1971), particularly pp.17, 98-101.
(2) In 1972 in Malaysia, gravel pumping employed 72 workers per ton of output and dredging 39 (Thoburn, 1977a, p.41).
(3) This information on Billiton is from an account in Dutch of the company's early history by Mollema (1922).
(4) This information is from a most interesting paper by Jackson (1969) in which the history of 18th century Bangka is researched from a rich variety of sources.
(5) See Wong (1965, pp.163-167 and 227-230) for further details. Also Allen and Donnithorne (1954, pp.158-59).
(6) Figures from Malaya are from Thoburn (1977b, p.87) and for Thailand from Ingram (1971, p.100).
(7) These details are from Klein (1965).
(8) These shares are based on the quotas allotted under the 1930's international control scheme (Klein, 1965, p.21).
(9) See especially Geddes' (1972) biography of Patiño, chapters ten, twelve and sixteen.

(10) For an exhaustive account of ownership and control in Malaya in the early post-war period see Putucheary (1960).
(11) According to Klein (1965, p.19) when CTS was formed Patiño bought out National Lead's holding in Williams Harvey.
(12) See Geddes (1972, chapter eighteen) for further details.
(13) See Yip (1969) chapter seven for a discussion of the schemes.
(14) Calculated from Raphael Zorn (1976). This source did not list output figures for some of the less important LTC companies, but it is known that the combined LTC-CC output share is about 27 per cent (TI, November 1977). Since dredging and open cast (dry)mining (mainly under CC control) in 1975 produced about 35 per cent of Malaysian output, this means that CC and LTC account for over two-thirds of foreign company output.
(15) See Lanning and Mueller (1979, p.278) for further discussion.
(16) Details from Lanning and Mueller (1979, especially chapters twelve and nineteen), who are especially useful on Anglo's methods of securing effective control at minimum cost via minority shareholdings.
(17) It is not clear whether control over such production could be maintained after it had passed through the smelter.
(18) Amalgamated Tin Mines of Nigeria (which produced about 45 per cent of Nigerian tin output in 1973 and in which LTC owned 30 per cent) accounted for 2.4 per cent of LTC assets, and its Thai companies 7.9 per cent.
(19) The total number of mines in 1976 is given as 538 in Mineral Statistics of Thailand, but only 4 suction boats are listed. Since there have been probably several thousand suction boats at work on a quasi-illegal basis, the importance of small scale operations in total numbers is thus understated.
(20) According to the Malaysian Department of Mines (BSMI, 1975) 'European' producers accounted for 54.9 per cent of open cast output and 82.5 per cent of underground mining output, but these two methods produced only 3.9 per cent and 2.9 per cent respectively of Malaysian tin output. In dredging and gravel pumping the 'European' shares of production were 92.7 per cent and 11.9 per cent respectively. The basis for tin definition of production as 'European' or 'Asian' is not given in the statistics, but it probably follows other official usage in referring to whether over 50 per cent equity ownership is owned by Europeans or Asians.
(21) Thoburn (1977a, p.43) shows that in 1974 the rate of return on a dredging investment was higher than a gravel pump operation at the then price of about M$1000 a picul, but it would have been lower than the gravel pumping rate of return at a price 25 per cent higher. At a 15 per cent discount rate, the net present values also switched in favour of gravel pumping. See also Thoburn (1978b). Chapter six shows that this phenomenon continues to the present.
(22) This is based on a series for 1950 to 1974 by Bird (1978, p.35), who uses the OECD GDP deflator. Engel and Allen (1979, p.113) also construct a real price index using the UN index of manufactured export prices as a deflator, with 1971 as the base, and covering the period 1956-78.
(23) The All Malaya Chinese Mining Association in a submission to the Malaysian Ministry of Finance in 1978 estimated that 80 per cent of its members were working on tailings or previously dredged land.
(24) Output in 1972 was 76,830 tonnes compared with 57,767 tons in 1950. There was however sharp export control under the ITA 1957-60 and lesser control in 1968-69, 1973 and 1975.

(25) The TEMCO dredge, delivered in 1971, has a capacity of about 240,000 cubic metres per month. The Aokam 3 sea dredge was almost certainly the only other new dredge purchased since the war. Its capacity is 420,000 cubic yards a month, (Rachan, Pow and Champion, 1969), about half the capacity of the latest Malaysian dredges.

(26) In 1974 offshore dredges produced 22 per cent and inland dredges 10 per cent of Thai tin output. In 1976 these proportions had fallen to 7 per cent and 9 per cent, while suction boats (shown in Department of Mineral Resources statistics for the first time that year) produced nearly 23 per cent of total tin output.

(27) Note that while LTC ownership in the other 'TASK' companies was substantial that in Aokam was only 4.6 per cent. Also, while the 1978 Annual Reports for Southern Kinta and Tongkah Harbour record Anglo-Oriental's management services for over 40 years (the Kamunting date is not given) Anglo-Oriental acted for Aokam only from 1967. In fact 29 per cent of Aokam was owned (in 1977) by Tronoh, a part of the Associated Mines group in Malaysia controlled by Charter Consolidated.

(28) In 1977-78 according to its Annual Report, Southern Kinta produced 64 per cent of its output in Malaysia and the rest in Thailand. Kamunting ceased its operations in Thailand in 1975.

(29) Faber Merlin has also had links with the Pahang Consolidated Corporation Ltd., the largest underground mine in Malaysia. In 1976 Faber Merlin owned 25 per cent of PCCL (Raphael Zorn, 1976).

(30) According to Fairmont State Ltd.'s Annual Report in 1978 the group also had property investment of a value equal to just under half of its total assets.

(31) See Saint Piran Ltd., AR 1978. The share listing of St. Piran was suspended on the London Stock Exchange in November 1979 because of certain alleged irregularities, and further details about its operations are difficult to obtain (The Guardian, London, 6.11.79 and 2.4.80).

(32) Fox (1974 p.78) notes that in the mid-1960's 'New management and fresh capital were pumped into South Crofty from the Siamese Tin Syndicate (now renamed the St. Piran Mining Co.).

(33) This statement is based on output figures from Zorn and Leigh-Hunt (1974) for 1971-73.

(34) Details from interview in Brussels, June 1978.

(35) Ownership details for all the major Australian tin mines are given in McKern (1976, pp.238-39). Company output figures are given monthly in TI.

(36) The acquisition was made via Anglo American's subsidiary, De Beers, the diamond company.

(37) AMC now holds 10 per cent in Malayan Tin Dredging, Southern Kinta, and Southern Malayan Tin Dredging, and 9.6 per cent in Amalgamated Tin Mines of Nigeria.

(38) According to the STC's 1978 annual report it had holdings in the following tin mining companies: Galian Jinjang (30 per cent), Hongkong Tin (29 per cent), Killinghill Tin (39 per cent), Kinta Kellas Tin Dredging (49 per cent), Petaling Tin (21 per cent) and Straits Tinfields (100 per cent). Of these, only Killinghill, Kinta Kellas and Petaling are in the TI list of major Malaysian tin companies, and they produced together approximately 2000 tonnes of tin-in-concentrates, about 3 per cent of Malaysian tin output. The 1977 Tronoh annual report listed a 17 per cent interest held by the STC in Tronoh (owned by Charter Consolidated), but this is not mentioned in the 1977 STC annual report. Hongkong Tin, Killinghill and Petaling are in the Osborne and Chappel group.

# 5 Policies towards Existing Production

In Malaysia and Thailand taxation has been used to secure for the host government a substantial share in the resource rents generated by tin, and there has been little difference between the tax regimes facing local companies and those facing foreign firms. The use of taxation to change the retained value from current production is discussed in section 4 which also assesses other local gains from tin production in the main exporting countries. Issues concerning the effectiveness of taxes in creaming off resource rents, and the likelihood of these tax levels discouraging foreign investment, are deferred until the next chapter. In addition to changing taxes, both Thailand and Malaysia in the 1970's have sought directly to increase government ownership and control. Malaysia has acquired control of the London Tin Corporation in a stock exchange takeover and has made foreign companies restructure their equity ownership. Thailand has refused to renew several foreign mining leases and has forced the restructuring of equity, as have the minor producers Zaire and Nigeria. The Thai and Malaysian actions are set out in section 2 and that of the minor producers in section 3. The chapter begins by looking at the effects of outright nationalisation, which has occurred in Bolivia and Indonesia.

## 5.1 THE BOLIVIAN AND INDONESIAN NATIONALISATIONS

In 1952 the 'big three' mining groups of Patiño, Hochschild and Aramayo in Bolivia were nationalised by the new Nationalist Revolutionary Movement (MNR) government, a regime which had seized power with the support of city dwellers and militant, armed mineworkers. The national mining corporation, Comibol (Corporacion Minera de Bolivia), formed to work these mines then controlled about three-quarters of the country's tin output. In Indonesia after the war the Billiton company had been given a five year contract to run the Bangka and Billiton island mines. The contract for Bangka was not renewed in 1953, and in 1958 the rights of Billiton on Billiton island and Singkep were taken over by the government (Fox, 1974, p.35). After various methods of administration, a State enterprise P. N. Timah (now P. T. Timah) was established in 1968 to run the country's entire tin mining operations.

It is easy to present the early effects of the Bolivian nationalisation as a catalogue of disasters. Comibol's tin production fell steadily from over 27,000 tonnes in 1952 to just over 15,000 in 1960. Its workforce grew from 24,000 in 1951 to 36,558 in 1956 and by the end of the 1950's labour productivity had fallen by about 60 per cent (1). The rise in the workforce undoubtedly represented the power of the mineworkers union and a system where workers were able to take an active part in controlling management decisions. The fall in output is partly a reflection of maladministration, but it also reflects an almost total lack of reinvestment in the previous twenty years (2) and a failure to discover new deposits during that period; with the result that ore grades declined. However, according to the estimates of a former general manager of Comibol, the corporation made an average annual profit of about \$10 million (on a turnover of under \$100 million) during the early years after nationalisation (Zondag, 1965, pp. 84-85). Revenue from Comibol was creamed off, principally by a system of multiple exchange rates whereby the corporation had to sell its export proceeds to the authorities at an artificially low price, in order to finance development elsewhere, especially in the petroleum sector and in agriculture.

The decline in Comibol's production must be set against a background of wider dislocation, partly caused by the mineral sector's own decline and partly as the short term consequence of land reform. In the five years after tin nationalisation GDP in Bolivia fell 20 per cent per capita and by 1956 the price level had risen by over 2,000 per cent compared with 1951.

A stabilisation policy was pressed by the US government (which was aiding the new regime), and it used currency reform in 1956 and a variety of decontrolling measures (including freeing the exchange rate) to achieve its objectives. Comibol was freed from the implicit taxation of the multiple exchange rates (3), and an export royalty replaced the previous income and export taxes. Simultaneously the private medium size miners were relieved of a similar obligation to sell their ore to the Mining Bank, an institution set up in 1936, and nationalised in 1939, to aid smaller miners vis-à-vis the 'big three' and large ore-buyers (Geddes, 1972, p.347).

The potentially beneficial effects of the 1956 stabilisation for Comibol were offset by a sharp decline in the tin price in the late 1950's, as Soviet tin exports depressed the world markets, the USA ceased stockpiling tin and the ITC introduced export control. Overmanning and lack of investment - only \$3 million in the whole of the 1950's - made for a continuing decline in output, only arrested after 1961 with a large influx of foreign capital under the so-called Triangular Operation for the Rehabilitation of the Nationalised Mines (Zondag, 1966, p.230). Under the Operation over \$37 million was provided by the US and German governments and the Inter-American Development Bank as loans for new exploration, re-equipment and to aid the reduction of surplus labour. Credits were also secured from Argentina and from large ore buyers (Zondag, 1965, chapter nineteen). In addition, the MNR government having strengthened its power base among the peasants (and agriculture employed the bulk of the Bolivian workforce) felt able to take a harder line about the mineworkers, a line strengthened after the takeover of power by a military government in 1964. During the late 1950's Comibol had made losses of about \$10 million per annum, though from 1957 to 1963 it paid no taxes on its exports.

The decline in Comibol's output after nationalisation in Bolivia is closely paralleled by the experience of the Indonesian takeover of the tin industry. The taking over of Billiton's remaining leases was part of a general expulsion of Dutch interests in 1957-58, and there was an outflow of Dutch personnel, who had been important in running the tin mines (4). Indonesia's tin exports fell from about 23,000 tonnes in 1960 to under 13,000 in 1966. These years were a time of general economic dislocation under Sukarno's 'guided economy'. In 1965, just before a military coup established General Suharto as the new president, inflation was running at 650 per cent per annum. An inappropriately high exchange was maintained, reducing the industry's profits in a situation where it had little influence on price, the divergence between the official and the black market exchange rate being about 500 per cent (Mackie, 1971, p.64). A serious problem too was the previous lack of investment. Although six new dredges had been commissioned between 1947 and 1948 the bulk of the dredge fleet was of an advanced age (P. N. Timah, 1972). In 1966 a new dredge was commissioned to work offshore at Bangka, and a rehabilitation programme was begun with Dutch aid following the military coup (Fox, 1974, p.36). A state tin mining enterprise, Perusahaan Negara Tambang Timah (P. N. Timah) took over the running of the mines from the army. After a trough in output of 12-13000 tonnes in 1966-67, during which time many Chinese workers were expelled from the islands, production rose steadily, reaching over 25000 tonnes by the mid-1970's.

Much has been made in the mining press and elsewhere about Comibol's losses. In 1978 it was reported to have lost \$60 million (on a turnover of about \$270 million) (5) and had an accumulated deficit of over \$150 million. It is possible, of course, for a company to run at an accounting loss and yet to yield net social benefits (6) (for example, if shadow wages are below market wages - see chapter two), but Comibol's losses often appear as an argument against nationalisation. In fact, in the 1970's in the years for which statistics are readily available (i.e. for 1971 to 1978) (7) annual average price always yielded a before tax profit, and it is royalty payments (a tax on output largely regardless of profitability) which have creamed off the bulk of the surplus and led to losses in 1976 and 1978 (8). In 1978 royalty payments turned a 22 per cent profit on turnover into a 2 per cent loss, and this was a figure which included the underground operations of the more profitable medium mining sector. In 1976 royalty payments were equivalent to 21 per cent of the average LME price (and taxes to a further 8 per cent), and in their absence Comibol would have made an average 18 per cent profit on turnover.

Heavy taxation of a state enterprise is not in itself necessarily undesirable. The state may well put the profits to better use than the enterprise itself, but in Comibol's case it is probable that opportunities for re-equipment and exploration have been lost. The Harvard team which in the mid-1970's advised the Bolivian government on its mining taxation policy (see Gillis, 1978) concluded that over the ten years to 1975 only about \$10 million had been spent on exploration (9) and only one significant new deposit had been found. Gillis and his associates also stressed a lack of fully effective management control, and noted the trade off between higher accounting profitability and the social responsibility undertaken in employing a larger labour force than technically necessary. Since then various new projects have been announced, including the National Mineral Exploration Fund set up in 1979 with International Development Agency finance of over \$35 million (_TI_, August 1979) and a dozen or so new projects (mostly Comibol's) were said to be under consideration, with outside finance of \$300 million being sought from the World Bank and other agencies. Yet in 1979 the resigning general manager of Comibol complained about the outdated equipment from which the corporation suffered and described Comibol as being on the 'verge of bankruptcy' (_TI_, April and October 1979).

In addition to Comibol's activities, roughly 25 per cent of the country's tin output is produced by privately owned mines. These are officially classified into a 'medium' and a 'small' scale sector accounting for about 10 per cent and 15 per cent of output respectively (10). The medium sector, which consists of nearly thirty firms, employs about 8,000 workers and all of its activities are underground except for the operation of two dredges. According to Gillis (1978, p.32) (11) only two firms were wholly foreign owned, and four or five had minority foreign participation, though these figures are not broken down to allow the importance of foreign ownership in tin production to be assessed (since some medium mines are important producers of tungsten and antimony too), and six mines in fact produce 70 per cent of the sector's tin output (12). Fox (1974, p.66) estimated that in 1966 40 per cent of the sector's output came from wholly Bolivian owned and under 10 per cent from foreign owned mines, Chilean and American capital being then of some importance. Small mines employed a further 3,000-4,000 workers and cooperatives about another 20,000, though the latter's output is very small, and much of this small scale mining is on a seasonal basis.

In 1966-69 some reinvestment occurred in medium mines resulting in updating of their equipment, but little new investment had taken place since then when the Harvard team visited Bolivia. Tin International (October 1979) reported investment programmes of $17 million and $22 million in 1977-78 but the medium mines were complaining strongly about the tax system based on royalties (which make no allowance for production cost differences) and they had tended to reinvest in agricultural ventures (where tax provisions were more favourable) rather than in mining. Medium mines, however, in many cases have lower costs than Comibol, basically because of higher labour productivity.

In recent years P. T. Timah has embarked on a programme of expansion based on seagoing dredges. The Banka 2 dredge, the first new one since 1966, started operations in 1978, and contracts were signed in 1978 and 1979 for two similar dredges for Belitung and Singkep. All three new dredges are large capacity, each to produce between 1000 and 2000 tonnes per year, and they form an important part of the planned increase to 35,000 tonnes by 1983 under the third five year plan, Repelita III (1979-83) (13). Equally important is a gradual replacement of the existing dredge fleet, 65 per cent of the capacity of which, prior to the commission of Banka 2, was of pre-war construction (14). A great deal of rehabilitation of the older dredges has been undertaken, however, including the conversion of all the steam driven dredges to electric or diesel-electric power. Rehabilitation has been internally financed by Timah, and taxation at the end of the 1960's was reduced to allow this. According to a Tin International special feature on Indonesia (January and February 1972), taxation stood at 75 per cent of gross proceeds in 1967 and was reduced to 15 per cent in 1968 and 10 per cent in 1970. This presumably refers to the export tax, and Table 5.1 shows that this actually took 12.7 per cent of export revenue in 1968, supplemented by a small income tax payment. In 1968 and 1969 too Timah was allowed to use part of profit before tax to finance rehabilitation expenditure (15), the allowance amounting in 1969 to 70 per cent of gross profits after export tax (P. N. Timah, Annual Financial Report 1969). Unlike Comibol, Timah's basically more profitable operation and lower tax levels have allowed it to continue to finance its rehabilitation and new investment internally. The change in corporate status to P. T. Tambang Timah Persero in 1976 gives it the power

TABLE 5.1

PERCENTAGE DISTRIBUTION OF GROSS SALES REVENUE, P. T. TIMAH, 1968-77

| | GROSS SALES REVENUE | OTHER REVENUE | PROFIT BEFORE INCOME TAX AND EXPORT TAX | DEPRECIATION PROVISION | EXPORT TAX |
|---|---|---|---|---|---|
| 1968 | 100.0 | 12.8 | 44.0 | 15.4 | 12.7 |
| 1969 | 100.0 | 6.7 | 31.5 | 18.5 | 12.9 |
| 1970 | 100.0 | 6.3 | 35.4 | 9.3 | 10.4 |
| 1971 | 100.0 | 15.6 | 36.6 | 12.2 | 9.6 |
| 1972 | 100.0 | 2.6 | 24.0 | n.a. | 9.6 |
| 1973 | 100.0 | n.a. | 25.4 | 20.1 | 8.8 |
| 1974 | 100.0 | n.a. | 24.5 | 9.6 | 8.8 |
| 1975 | 100.0 | n.a. | 20.1 | 4.8 | 10.4 |
| 1976 | 100.0 | n.a. | 28.1 | 3.7 | 8.5 |
| 1977 | 100.0 | 1.6 | 38.2 | 5.1 | 9.7 |

| | INCOME TAX | INVESTMENT EXPENDITURE | OUTPUT (TONNES) | TIN PRICE (M$ PER PICUL) |
|---|---|---|---|---|
| 1968 | 5.7 | 6.0 | 16898 | 565 |
| 1969 | - | 8.6 | 16541 | 626 |
| 1970 | 8.7 | 17.1 | 19092 | 665 |
| 1971 | 10.9 | 15.4 | 19767 | 632 |
| 1972 | 6.4 | n.a. | 21766 | 627 |
| 1973 | 8.4 | 28.4 | 22490 | 686 |
| 1974 | 7.6 | 10.0 | 25023 | 1137 |
| 1975 | 5.1 | 13.1 | 24391 | 963 |
| 1976 | 9.4 | 13.2 | 24258 | 1147 |
| 1977 | 12.7 | 3.3 | 22103 | 1588 |

Sources and Notes

(1) 1968-72 distribution figures from Development Consult Inc. PT (1974 A-206), 1973-77 from P. T. Timah, Financial Report 1977 (in Indonesian). Prices from TI. Output figures from P. N. Timah (1972) P. T. Timah (1976) and Indonesian Mining Yearbook (Department of Mines and Energy, 1977).

(2) Other Revenue in the 1968-72 figures includes proceeds from the Peltim smelter and Peltim sea transportation as well as interest, etc.

(3) Depreciation is included in costs prior to calculation of Profit.

(4) Investment is taken as the figure for additional fixed assets (penambahan tetap) for 1973-77.

to raise money on the open market, the change from 'P. N.' to 'P. T. Persero' indicating that it is no longer a State enterprise but a private limited company (albeit with the government as sole share owner). After tax profit in 1977 was equivalent to US $39 million (on a turnover of $250 million), which is enough by itself to have financed an entire new sea dredge project (the Singkep I dredge project signed in 1979 was for $31 million, for example). In practice the shareholders (the government) decide on the distribution of profits (16) but investment plans are submitted to the investment coordinating body, BKPM (17) which allows offset against tax (18). Table 5.1 shows that investment expenditure has been substantial in the 1970's.

The use of a light export tax (currently 10 per cent of the value of tin metal sales) and a substantial tax on Timah's profit (at 45 per cent) is in accordance with the generally accepted principles of mining taxation set out in chapter two. In theory, tax could be levied entirely on profits though in practice this gives greater scope for evasion or (more likely in a State company) the inflation of costs by, for example, excessive staff benefits. There is no provision for taxes to rise if the tin price does so (which is in contrast to the current third generation contract applied to new foreign mining investment in Indonesia, discussed below), but undoubtedly the system is considerably more efficient than the Bolivian export royalty system in the sense of allowing adequate resources for new and replacement investment and exploration while also generating substantial tax revenue. One effect of investment and rehabilitation has been a considerable rise in labour productivity. Output per worker per year has risen from 0.5 tonnes in 1966 to roughly 0.9 in 1975 (19) and one tonne per worker is the current target (<u>TI</u>, April 1979). This compares with 1.6 tonnes in Malaysia in 1975 where grades are <u>lower</u> (<u>BSMI</u>, Malaysia, 1975). However, the figures are not strictly comparable since Timah is far more vertically integrated than the Malaysian tin industry, supplying its own power and repair facilities whereas the Malaysian mines buy such services from outside the industry and workers in the supplying industries are not included in the Malaysian statistics (20). Timah however, regards itself as having social obligations to employ a workforce larger than necessary on strictly commercial grounds. It is also worth noting that mining costs per tonne of concentrates differ substantially between the tin islands. Bangka produced 69 per cent of 1977 output but accounted for only 53 per cent of Timah's mining costs (21), the difference being primarily a reflection of Bangka's higher ore grades.

Although tin is Indonesia's largest mineral export after petroleum, P. T. Timah occupies a far less central position in the mining economy than does Comibol. Whereas Comibol also accounts for large shares in Bolivian production of other minerals (excluding oil) (22), Timah has had the power to diversify only since its change of status in 1976 and has moved only to a small extent into kaolin, bricks and other minor items. Other state enterprises in Indonesia are responsible for other mineral production - Pertamina for oil, P. N. Batubara for coal and P. T. Aneka Tambang for other minerals. Timah has recently started a joint venture with Rio Tinto Zinc for exploration for minerals (apparently copper in particular) in Sumatra, however.

Though Timah produces about 90 per cent of Indonesia's tin output, not all of its production is from its own facilities. Gravel pumps produce nearly half of total output and slightly more than half of these mines are operated by contractors. Although in the small project of Bankinang in Sumatra the contractor supplies his own equipment and fuel, and sells the ore to Timah at a fixed price, in Bangka and Belitung the contractors normally supply only labour. Presumably the system gives the corporation some flexibility in altering the effective size of its labour force. It may also be a means of using the mining expertise of the Chinese community who are excluded from direct employment (23). Since the payments to contractors appear not to vary, at least in the short term, with the tin price, there may be some incentive to smuggling at times of high prices. Early in 1979 there was a major drive against smuggling from Bangka to Singapore.

Foreign investment in tin has been allowed since the change of regime in the 1960's under the 1967 Foreign Capital Investment Law. Only three multinationals have entered the industry (24) (25). Billiton re-entered Indonesia in 1968 with a 40 year contract to prospect and mine offshore in the Riau archepelago near Singkep. In 1979 it started mining with the world's largest tin dredge, the project costs being about US $65 million. As noted elsewhere in this study, Billiton is now wholly owned by Royal Dutch Shell, though its Indonesian subsidiary is 10 per cent owned by Timah, and it must offer 25 per cent of its equity to Indonesians over ten years. Broken Hill Proprietary, the second foreign company, whose contract dates from 1971, has re-opened an old lode - the only underground mine in the industry - on Belitung island. Broken Hill's parent company, of the same name, is one of the three largest mining companies in Australia (McKern, 1976, p.13) (and it also manufactures tinplate - TI, January 1980) but the Indonesian operation is said to have yielded little profit and Timah has not exercised its option to take up an equity stake. Whereas BHP and Billiton have offered specialised mining expertise in areas where Timah to some extent still lacks it, the third venture, Koba Tin, rather surprisingly operates a mainly gravel pump operation of a sort similar to Timah's. It is apparently extremely profitable and Timah holds 25 per cent of its equity with the option of purchasing up to 45 per cent at 2 per cent per annum. The foreign owners are the Australian companies Colonial Sugar Refining and Blue Metal Industries, the former being an active mining company.

All three multinationals operate under 'second generation' contracts, basically similar to those in other mining (26) over the period 1968-76 (27). Companies pay a royalty on tin output of 8-10 per cent, though not the export tax (28). In addition they pay the 45 per cent company income tax (40 per cent for the first ten years of operation) and there are various minor local taxes and sometimes import duties. Tin is now said to be closed to foreign investment, and in any case any new company would be subject to the more stringent conditions of a 'third generation' of contracts (29).

Before the Indonesian and Bolivian nationalisations took place, the bulk of each country's tin concentrates were shipped overseas for smelting. In the Indonesian case they were sent to the Netherlands until 1959 and then to Malaysia until confrontation in 1963 (Fryer and Jackson, 1977, p.195) though before the war there had been some local smelting. From Bolivia concentrates were mainly sent to Europe, where the Patino organisation owned the largest smelter, Williams Harvey. Since nationalisation both countries have integrated forward into smelting and the Indonesians have established a fully integrated organisation through to marketing of the metal.

The smelting of alluvial tin concentrates, in Malaysia at least, has proved to be an extremely profitable occupation (30) and this seems to be true of Thailand too (31). In Indonesia, producing basically similar concentrates, considerable difficulties were experienced in the early stages of establishing a smelter, which was designed and built over the period 1961 to 1967 by a German company. The major problem, apparently, was that the smelter was of an essentially experimental design - using rotary furnaces instead of the more usual stationary ones - and by 1970 had achieved less than two-thirds of rated capacity. Additional, more conventional capacity designed in Denmark and built by a Singapore firm, installed during 1973-75 has given Indonesia self-sufficiency; all concentrates now being locally smelted except for those from the single (foreign owned) underground mine, which are low grade and are sent to Malaysia. In 1979 the smelting unit was building a further furnace to raise total capacity to 39,000 tonnes per year (32) to deal with increases in future output.

According to Timah's director of marketing in 1973, (ITC, 1974, p.15) the smelting operations already 'made money', though in 1978 costs appeared to be higher than the charges levied by Malaysian smelters (33). Whether or not some cost saving could be achieved by continuing to ship concentrates to Malaysia, however, smelting does permit Timah to market vitually the whole of Indonesia's output, much of it on long term contract. Agents are used in the US, France, Italy and the UK; and to Taiwan, South Korea, Eastern Europe and the Phillipines there are direct sales to consumers. In Germany and Japan there are sole buyers. For example, having been unable to penetrate the Japanese market, Timah sells solely to Mitsubishi, who resells.

The establishment of local smelting capacity has taken considerably longer in Bolivia, and as of 1979 40 per cent was still being sent abroad. In the very recent past the bulk of the ores shipped overseas have gone either to Capper Pass (owned by Rio Tinto Zinc) in England or to the Texas City Smelter in the USA. Smelting the complex hard-rock ores of Bolivia poses more problems than are found with the purer alluvial concentrates of Asia. In Asia most mines concentrate their tin bearing ore to about 70-75 per cent purity. In Bolivia the purity has fallen from about 60 per cent in the 1940's to 35-50 per cent today, and many small mines deliver concentrate of below 20 per cent (Gillis, 1978, pp.34-36). The Bolivian smelting organisation ENAFBOL (Empresa Nacional de Fundiciones de Bolivia) is organisationally separate from Comibol and has been operating a smelter since 1971, built by the same German company which constructed the first Indonesian smelter. Costs were high in the beginning because infrastructure was built for a planned 20,000 tonnes per year capacity whereas only 6,000 tonnes was built initially, though it had risen to 11,000 by 1976 (Gillis, 1978, pp.34-35). The Enaf smelter can only handle the higher grade concentrates and in 1979 the building of a low grade smelter was being undertaken together with the expansion of the main smelter (at Vinto) to 20,000 tonnes to give Bolivia self-sufficiency (<u>TI</u>, April 1979).

Whereas the Indonesians appear satisfied with the performance of their smelter the outgoing manager of Comibol complained in 1979 that he could make another $20 million a year if freed from the obligation to sell his concentrates to Enaf (TI, October 1979). This alleged loss represents 7 per cent of Comibol's turnover, while in Malaysia smelting costs normally less than 2 per cent of the value of output. The Bolivian operation has suffered from lack of suitable fuel and lack of higher grade concentrates to achieve the suitable mix than overseas smelters have obtained in the past (Zondag, 1965, p.97). Nor have the Bolivians successfully entered marketing in the Indonesian fashion. In 1973 some 70 per cent of medium mines sales were directly to foreign smelters though some were to the mining bank BAMIN (Banco Minero) which also buys the output of the small mines. According to Widyono (1977, pp.33-36) international metal trading companies are still very active in Bolivia, their commission being up to 20 per cent, and the main firms being Phillip Brothers, Tennants, Metal Traders, and a subsidiary of the Amalgamated Metal Corporation. However, selling metal increasingly rather than concentrates improves the Bolivian's bargaining position by somewhat increasing the range of potential buyers (34). The increased share of tin smelted in the producing countries has also meant the closure of many of the smelters in Europe important in colonial times - thus Williams Harvey in England and the smelter at Arnhem have gone into liquidation.

## 5.2 THE APPROACHES OF THAILAND AND MALAYSIA

The Malaysian government's takeover of the London Tin Corporation has been in line with the 'New Economic Policy' initiated in the Second Malaysia Plan (1971-75) after serious racial communal riots in 1969 had demonstrated a deep dissatisfaction among Malays with their economic position. Under the Outline Perspective Plan to 1990 it is intended that the economy will be restructured so that asset ownership will reflect more accurately the shares in population of the various ethnic groups. It is targetted that foreigners should own only 30 per cent of the country's assets, Malays 30 per cent, and other Malaysians 40 per cent. The Malaysian portions will then represent more closely than at present the shares of each ethnic groups in total population - which in Peninsular Malaysia is 53 per cent Malay, 35 per cent Chinese and 10 per cent Indian.

The LTC takeover was achieved in a complicated series of manoeuvres involving Haw Par Brother International in Singapore, Charter Consolidated and, for a time, the British property company Slater Walker. Control of Haw Par, a firm which owes its growth to the famous Tiger Balm medicines, and which had also a variety of banking and newspaper interests, was acquired by Slater Walker in 1971 as part of an expansion into Far Eastern markets. 29 per cent of the equity of the LTC was bought by Haw Par from four London institutions in 1973 (Slater, 1978, p.188), following a series of other acquisitions in Hongkong and elsewhere. In 1974, when Slater Walker sold Haw Par as part of a general move into cash in difficult stock market conditions (Raw, 1978, p.383), a substantial part (13 per cent) of Haw Par was bought by Charter. Very little information is available about Haw Par's motivation in acquiring a stake in LTC (35), though certainly the LTC at the time was extremely cash-rich. Malaysian control of the LTC was reached in two stages. First, and apparently on the initiative of the then head of Haw Par, an approach was made to the Malaysian national corporation, Pernas, who acquired by 1975 21 per cent of LTC shares (36). This initial attempt was prevented by the Singapore stock exchange, which required Haw Par to bid for all of the LTC shares, which it was unable to do. Subsequently, in 1976, Pernas itself bid for the whole of LTC's equity on the London market, and an exchange of shares was arranged with Charter as Pernas' new partner. Under this, Pernas, through its subsidiary New Tradewinds, acquired Charter's Malaysian tin interests - the Associated Mines/Tronoh group - in exchange for Charter receiving 29 per cent of New Tradewinds' shares (37). New Tradewinds, shortly afterwards renamed the Malaysia Mining Corporation, controls as a result over 25 per cent of Malaysian output, which in turn is about two-thirds of the output of companies of foreign origin. MMC is managed by Pernas Charter Management, in which Charter and Pernas hold 50 per cent each. The formation of MMC leaves outside the government's control only the Osborne and Chappel group (in whose companies the smelter, the Straits Trading Company, has interests) and individual companies such as Conzinc Rio Tinto and Pacific Tin.

The formation of the Malaysia Mining Corporation has not been a means of achieving total ownership, since LTC control (as mentioned in the previous chapter) was secured through minority holdings, ownership of the other shares being widely dispersed. Research on local share ownership in dredging companies in the 1960's and 1970's (38) suggested that Malaysians owned about a third of the equity and in 1977 the Far Eastern Economic Review (1 April, 1977) estimated the overall Malaysian ownership of MMC companies at 45 per cent. Details about the current local ownership of its companies could not be obtained from MMC (39), though some are claimed to have very high local ownership (40). Adding the known LTC and CC ownership proportions to 1970 figures for local ownership from Thoburn (1977b, p.280) suggests a combined local ownership figure of nearer 60 per cent for the MMC companies as a whole, though the 1970 figures were complicated by nominee holdings and some 'local' shareholders may well have been Singaporean.

For Pernas, established under the New Economic Policy to hold assets in trust for indigenous Malays, the acquisition of the LTC and CC tin interests has been an important means of moving its previously loss-making group into profitability. In the 1977-78 financial year Pernas recorded profits for the first time. It is worth asking, though, whether the M$170 million paid by Pernas for the LTC was reasonable. The market value of the LTC companies in 1970 was slightly in excess of M$300 of which the LTC owned just over 30 per cent.(41)The nominal 197 pence per share paid for the LTC by Pernas compares with a 1971 price range of 95-135 pence, but considerable rises in all tin shares occurred following the 1973-74 price boom and it is difficult to distinguish the effects of Haw Par and Pernas purchases. Fortunately for Pernas the takeover antedated the really sharp price rises of the late 1970's, and all in all it seems likely that the price paid was reasonable. This is certainly the view argued by Pernas' chairman, Tengku Rezaleigh (in Tin, April 1980) who describes how a 25 per cent fall in the share price took place between the first meeting between Pernas and Slater Walker in 1972 and Pernas' first purchase of 20 per cent in the LTC in 1975. Pernas gained further from the fact that the British authorities allowed the (43 per cent) dollar premium to be included in the bid price when the remainder of the LTC's shares were purchased (which reduced the effective price per share from 197 pence to 137) and from the fact that the Malaysian ringgit had appreciated against sterling over the 1972-75 period.

Malaysian ownership restructuring in tin was part of a well articulated general policy towards foreign investment, the terms of which were made known well in advance. The initial moves towards restructuring in Thailand were a haphazard accompaniment to a change of government. In 1973 the military regime of Thanom and Praphat was overthrown in the context of student uprisings, and in the south of the country there was agitation against foreign ownership of natural resources. The most striking outcome was the revocation of the TEMCO lease in 1975, which had been owned by Union Carbide and Billiton, on the grounds that it had been obtained improperly by UC from the earlier government. The revocation of the lease was preceded by an almost explosive growth of illegal mining by small suction boats, mentioned briefly in earlier chapters. According to the study by a team at Chulalongkorn University (1976) suction boat operations off the coast in Phuket province started in the mid-1960's, the operating methods gradually becoming more sophisticated. According to miners in Phuket (42) both TEMCO and other foreign dredging companies had experienced difficulties with illegal miners before, but up to 1973 the companies had been able to enlist the support of the local police to drive them off. After the change of government, the local administration, particularly in Phanggna province, was sympathetic to the small boat owners, whose operations on foreign owned leases were no longer checked.

Following the revocation of the TEMCO lease and a change back to military government in 1976, other measures have been taken both to decrease foreign ownership and to place it on a more certain basis. Existing mining companies must allow a majority Thai ownership 'at an appropriate time'. New offshore dredging ventures are allowed but must have majority local ownership from the beginning, and this must rise to 60 per cent within five years, whereas existing offshore dredging companies must immediately conform to the 60 per cent share (43). Aokam, which has been under the control of the Malaysia Mining Corporation, has reconstituted itself as a 60 per cent Thai owned Thai company - the Aow Kham Thai Company (Bangkok Post, 31 October 1978). Of the other members of the Malaysia Mining Corporation's TASK group in Thailand, Kamunting had its application for an inland lease refused and it left Thailand in 1975. Southern Kinta, particularly hard hit by illegal miners, shelved plans to construct a large capacity shallow digging dredge for offshore work, and its recent annual reports still complain of this problem. Tongkah Harbour, the fourth TASK company agreed at the end of 1979 to restructure to 60 per cent Thai ownership, but it was announced early in 1980 that the Thai government had not yet agreed to renew its lease (TI, March 1980). The other group of foreign origin, Fairmont State, which works inland, apparently has majority Thai ownership already (Interview with Fairmont State, Bangkok, October 1979).

Billiton Thailand Ltd., formed after the end of TEMCO and the withdrawal of Union Carbide, now works the old TEMCO lease as a contractor to the new leaseholder, the Offshore Mining Organisation, set up by the

Thai government in 1975. As of 1978 Billiton had four years of a five year contract to run, under which profits are shared. Owing to changes in export royalty conditions in 1977 and 1978 however, OMO was claiming it was losing money on the arrangement in 1978 (Bangkok Post, 27 October 1978), since OMO was responsible for paying the royalty. As a contractor to OMO, Billiton remains 100 per cent foreign owned (by Royal Dutch Shell) and is immune from the ownership restructuring. The only other exception is for hypothetical offshore dredging projects working below 200 feet, a depth not possible with present technology. Both OMO and Billiton continue to complain about the activities of suction boats, apparently backed by powerful local interests in southern Thailand. In particular the low recovery rate of the suction boats - claimed to be about 30 per cent - is said to spoil deposits, which would then need a more powerful dredge to extract the ore if the area were reworked. Undoubtedly the boats are 'highgrading', but it is difficult to see how, if 70 per cent of the ore is simply returned to the sea bottom, the deposit is effectively 'spoiled', given the very high grades available offshore. The returns to investment in suction boat mining are discussed in the next chapter, but it is worth saying here that they are very high. In fact, the Billiton-OMO venture is less troubled by suction boat operations, than is Southern Kinta, since the latter has a shallower lease. There are also areas in the Billiton lease where the seabed overburden is too deep for the boats' equipment to penetrate but which can be worked by Billiton's suction cutter.

Although formed to control the former TEMCO lease, OMO has begun to branch into wider mining activity. A dredge of its own has been ordered with a view to starting operations in 1980 (TI, September 1979). Setasup Karnrae, a local Thai-Chinese dredging company with various links with the Malaysian Chinese mining community, has acquired second-hand dredges also to work as a contractor to OMO, and in 1978 another large local mining company (also with Malaysian links) was building a suction dredge also to work as an OMO contractor. The Thai restructuring seems to have a more adverse effect on foreign investment than that in Malaysia, in spite of the fact that the maximum permissible foreign equity holding is smaller in Malaysia while the taxation levels are very similar. A consortium of Bethlehem Steel, Alcoa, Tronoh and Charter Consolidated was reported as pulling out of a joint offshore mining venture in 1978, and only one new project (a Thai-German joint venture) has been reported more recently. Southern Kinta is known to have abandoned plans for a shallow digging offshore dredge. Partly the TEMCO affair has damaged confidence in the legal position of foreign investors, and in spite of the change of government the small boats still poach on foreign companies' legally obtained leases, thus discouraging both exploration and new investment. Some inland mines in the south also have problems with communist insurgents and pay 'taxes' to them. The fact that Thai output has not fallen during the restructuring, except in 1975, is a reflection of the emergence of suction boat mining as the largest and fastest growing sector. The rapid rise in Thai output from 20,000 tonnes in 1974 to over 30,000 in 1978 is also a reflection of the boats being persuaded to sell their ore legally instead of smuggling it to Singapore. Phanggna province has set up a buying agency, subcontracted to various individuals (44). It is also worth noting that local miners, other than suction boat operators, seemed pessimistic about prospects when interviewed in 1978. Most reported they had diversified substantially in the last ten years, into property, plantations and finance. Common complaints were the level of taxation, official concern with environmental protection the miners considered excessive and a difficulty in obtaining bank finance.

In Malaysia the MMC takeover was accompanied by a large fall in output - from around 75000 tonnes at the beginning of the 1970's to a low of just under 59000 tonnes in 1977 - though the declining trend was halted in 1978. Undoubtedly the restructuring of ownership, under which foreign companies can hold normally only 30 per cent of the equity, has acted somewhat to discourage new projects. Complaints have also been made by all sectors of the industry in Malaysia about 'penal' taxation (45); and the effects of high tax rates and government equity participation are analysed in some detail in chapter six. Although dredging output failed to expand at a time of extremely high prices in the 1970's, the contraction of output is due to a decline in the gravel pumping sector. This has been partly due to a taxation structure which affects this sector particularly adversely but also due to difficulty in securing leases and finding new mining land.

The lack of prospecting in the 1970's and the consequent shortage of new land can partly be blamed on the New Economic Policy and on taxation, but for foreign companies especially, leases have been difficult to secure however strong was the company's willingness to invest. This problem is overwhelmingly the result of a conflict of interest between the federal government and the governments of the individual states. In the Malaysian federation the latter have the virtually exclusive right to grant new leases, yet they receive only 10 per cent of the tin export duty, none of the other taxes on mining, and only a small land premium. Thus their incentive to alienate mining land is slight. The two west coast states of Perak and Selangor produce about 90 per cent of the country's tin output, and Selangor has taken the lead in a tough policy to secure additional gains for the state government. Initially, even the Malaysia Mining Corporation had lease applications refused in Selangor, apparently on the grounds that its operating companies did not conform to the NEP ownership guidelines.

The vehicle for greater Selangor state government participation, since it does not have the power to tax, has become the Selangor State mining corporation, Kumpulan Perangsang Selangor. KPS's initial finance came from money paid by the federal government to Selangor in exchange for ceding Kuala Lumpur city as the federal capital, and since 1978 the declared state government policy is that no leases over 500 acres - effectively any dredging lease - will be granted to the private sector. With foreign companies, KPS will hold 70 per cent and with local companies 51 per cent (46), and these conditions apply to the renewal of existing leases too. In practice the guidelines seem to have been applied with some flexibility, although the distinction between 'local' and 'foreign' can sometimes be unclear. Conzinc Rio Tinto, RTZ's Malaysian subsidiary, has a joint venture in which there is a 51/19/30 per cent split between KPS, a local Malay company, and CRM. Pacific Tin, the only North American company in Malaysian tin mining (47) acquired a lease in 1979 for payment of 10 per cent tribute to KPS, in addition to which KPS acquired 10 per cent of Pacific's equity with an option up to 30 per cent. KPS has also entered mining directly through its own dredging company, Timah Langat, initially 80 per cent owned by KPS and now 56 per cent, the rest of the equity having been sold to local Malays. State Development Corporations in other states have also entered into joint ventures, though generally on less stringent terms than in Selangor. Perak has a 30 per cent stake in a local Chinese dredging project and in a large dredging venture where MMC holds 40 per cent and Tronoh mines directly holds another 30 per cent. Joint ventures with obviously foreign companies must conform to the '70/30' Foreign Investment Committee Guidelines, and CRM has a project in Perak on this basis, and in Pahang (48).

Selangor has also insisted on treating the Malaysia Mining Corporation Operating Companies as foreign and has stipulated a 70 per cent KPS stake even before leases have been renewed. In the huge new Kuala Langat deposit in Selangor where reserves are estimated as equivalent to about five years total Malaysian output, MMC again has been able to secure only a 30 per cent share; and this after protracted negotiations and the cancellation of a previous agreement (TI, December 1979).

In general the joint venture system involving state governments appears at last to have resolved the problem of leases for dredging, both for foreign companies and the MMC. Since there is significant foreign ownership in many of MMC's operating companies, the increased participation by state governments and Malay companies has resulted in greater local ownership, though the shares in Malay companies are not widely dispersed among the Malay community. MMC recently announced a major exploration programme, including offshore areas, whose cost (M$16 million) is equivalent to a complete dredging project. Certainly there is evidence that the remaining multinationals are willing to continue investing. For local Chinese miners the pressure to restructure in favour of Malays is more subtle. Officially even in Selangor there has been no change of policy towards local companies requiring lease renewal (or new leases) of under 500 acres, though in practice there may be some preference given to bumiputra companies for new leases. When interviewed in 1978 the Bumiputra Chamber of Mines had about 120 members - there are about 800 gravel pump mines in Malaysia - but roughly half had Chinese partners. Though, as the next chapter will show, tax structure is another factor adversely affecting the gravel pumping sector, the problem of increasing Malay participation while making use of Chinese expertise is acute. Malay participation in the mining labour force has increased to over a quarter, compared with about 5 per cent in the colonial past (Thoburn, 1977b, p.92), though some state government enterprises such as Timah Langat employ a wholly Malay workforce. Many mines prefer an ethnic mixture, however, since the mine can then be kept open through the country's many festivals.

In Thailand no moves to increase government ownership and control in smelting and marketing have occurred yet. The Thaisarco smelter remains 100 per cent owned by Billiton, which remains 100 per cent owned by Royal Dutch Shell, and Billiton markets Thaisarco's entire output. However, in January 1980 the Thai Offshore Mining Organisation was reported (by Tin International) to be negotiating an equity stake in Thaisarco and considering a joint marketing agency. Unlike Malaysia the smelter does not have its own network of dealers to buy from small miners and ore dealing is predominantly in the hands of the larger gravel pump miners together with the representatives of the provincial organisation which buys from the small boat operators. Some smuggling continues and the smelters in Singapore (a country which itself produces virtually no tin) are producing annually about 4000 tonnes of tin (TI, May 1980), mostly, it is thought, from Thailand and Indonesia. The establishment of Thaisarco in 1965 was supported by Thailand's Board of Investment with a prohibition on exports of unsmelted ore (which mostly had previously been sent to Malaysia) and a five year ban on other smelters being established. In 1976, permission was given by the Board for two new smelters to be established in the region near Bangkok (49), to deal with ore from the central and northern regions of the country, thus effecting considerable transport cost savings. Both are expected to be in production by 1980-81. The Thai Present Company is 90 per cent foreign owned but little other information is available about it. The Thai Pioneer Company, which is Thai owned, has links with the German metal dealing firm Metallgesellschaft which will supply smelting expertise and deal with marketing (50). Together the two companies will have a capacity of 12,000 tonnes per annum. In 1980 the planned location of the Thai Present smelter was switched to Phuket, where it will be in competition with Thaisacro (TI, May 1980).

No information is available about the profitability of the Thaisarco smelter, which charges according to the same scale as the smelters in Penang. Some additional profits will have been made from sales of tantalum, a byproduct which the miners cannot separate themselves and for which they are not paid, and Billiton (Interview, June 1978) claimed that it was smelting rather than marketing which was the profitable part of the operation. In Malaysia, where accounts for the two smelting companies are available, there is a long history of high profitability (Thoburn, 1977b, p.127).

Negotiations are proceeding in mid-1980 for the Malaysia Mining Corporation to acquire an equity share in each of the two smelters. Perhaps largely as a bargaining move, MMC has opened bids for the construction of a third smelter to handle its entire output. In any case MMC has dealt with its own marketing since 1978, when it started to toll smelt its output with the local smelters, and cooperate in marketing with Anglo-Chemical & Ore in London (TI, January and May 1980), a subsidiary of the large metal trading firm Phillips (also active in Bolivia). Previously all ore had been sold from Penang on an armslength basis, with the result that miners could not keep control of their ore through the smelter. This may well have been a factor too in the lack of integration backwards into tin mining by steel companies owning tinplate capacity. Bethlehem Steel for example, when interviewed in Bangkok in Autumn 1978, mentioned this difficulty in securing a captive source of tin.

The bypassing of the Penang market by MMC (and more recently by the Selangor state mining organisation KPS) has been described by the Malaysian Minister of Primary Industries as an anticipation of the establishment of a Kuala Lumpur Metal Exchange (TI, January 1979). The exchange, which will also trade in soft commodities, is due to start operations late in 1980, first with palm oil then with tin (51). The aim is to provide an opportunity for future trading in tin to supplement the physical activities of the Penang market, and to move a proportion of 'paper' trade in tin away from the New York market and the London Metal Exchange.

## 5.3 POLICIES AMONG MINOR PRODUCERS

Nigeria and Zaire have taken measures similar to those of Malaysia and Thailand to increase local ownership. Under the Nigerian Enterprises Promotion Decree first enacted in 1972 foreign companies by the end of 1978 were expected to have sold 60 per cent of their equity to Nigerians. The largest foreign company, at the time a subsidiary of the Malaysia Mining Corporation, (who had taken over from the London Tin Corporation) - Amalgamated Tin Mines of Nigeria - noted in its 1978 Report that its shares had been sold mainly to a state body, the Nigerian Mining Corporation. Shares were sold at a price decided by the Nigerian Capital Issues Committee. Similar sales have been made by other large companies but there have been complaints about the need to sell shares on a saturated market (TI, December 1978, February and December 1979). The Nigerian Mining Corporation also has taken a 21 per cent interest in the Makeri smelter (Engel and Allen, 1979, p.24). In Zaire, the Sominki company was formed in 1976 as an amalgamation of three smaller mining groups. According to Tin International (December 1979) 28 per cent of the equity is held by the government of Zaire. The only other foreign mining company in tin in Zaire, also of Belgian origin, Géomines, was reformed in 1966 into Zairétain, in which the Zaire government holds 50 per cent (52). In Ruanda, the Somira company was established in 1973 to include a number of the major tin companies, and is 49 per cent owned by the government (TI, November 1978). Australia requires 50 per cent local ownership but this is administered flexibly (Engel and Allen, 1979, p.153).

Taxation levels vary considerably between the minor producers as far as export duty is concerned, the rates for Nigeria being 16 per cent regardless of price, 2-4 per cent for Australia (varying with state) and zero for Zaire. Corporation tax is levied in those three countries at 45 per cent, 46 per cent and 50 per cent respectively. In the UK no export duty is payable and corporation tax varies from 42-52 per cent according to the size of profits (53).

In Zaire and Nigeria output has declined markedly in recent years. In 1978 Zaire's output was little more than half that of the late 1960's and that of Nigeria was little more than a third. Further research would be necessary to establish the reasons for these falls but it is known that Nigeria has experienced a great outflow of foreign personnel since its takeover. Engel and Allen (1979, p.24) attribute the Nigerian decline in part to falling ore grades and a lack of past exploration.

## 5.4 THE DISTRIBUTION OF GAINS FROM TIN PRODUCTION

In the discussion in chapter two of measuring the benefits to the host-country from mineral development, stress was laid on retained value, the country's share of final consumer value, and on wider development effects such as local market creation, income distribution changes and employment creation. Such measures can only be used by themselves to summarise mineral export development effects if tin export activity is 'marginal' to the economy (54). If it is not marginal one would expect the equilibrium exchange rate in particular to differ from what it would be in the absence of the mineral export activity, and therefore there will be benefits and costs in other sectors beyond those identified through the linkage analysis (55).

Of the four major tin producers, only in Bolivia is tin mining clearly non-marginal. There it generates 57 per cent of export income compared with 11 per cent in Malaysia, 7 per cent in Thailand and 2 per cent in Indonesia (1977 figures). This view is strengthened by a recent World Bank study (Nankani, 1979) (56), which argued that LDC 'mineral economies' have distinct characteristics, problems and prospects compared with other primary exporters, and which uses as the criteria for inclusion in that category that more than 10 per cent of GDP and over 40 per cent of export earnings (over an average of several years) should be generated by minerals. Of the tin producers only Bolivia qualifies (though Indonesia and Nigeria are included on account of their petroleum, as is also Zaire for its copper). Again, the distinct problems are thought to arise principally because of effects of export expansion on the exchange rate. Also, Nankani argues that high mining wages reinforce a dualistic economic structure of the sort described in section 2.2. In consequence, the first subsection here, which discusses non-marginal development effects, concentrates on Bolivia, though there is a brief account of tin mining and income distribution in the other main producers.

### 5.4.1. Exchange rate changes and other non-marginal effects

In chapter two (section 3) a discussion was given of the methods used by the Harvard team in Bolivia in 1973-74 to measure the mining sector's non-marginal effects (Gillis, 1978, pp.82-89). Here the empirical results of their method can be summarised. In the absence of the large non-fuel mining sector (predominantly tin mining) the exchange rate would need to depreciate. As a result of the depreciation the local currency price of non-mining exports would rise (57), and their supply would increase, while the local currency rise in import prices would reduce the quantity of imports purchased. The resulting change in welfare is the saving in resource costs resulting from the absence of mining activity, less the additional costs of increasing the supply of other exports, and less the welfare costs of the change in expenditure and fall in consumer surplus of the reduced import volume. Using linear demand and supply curves and making simple assumptions about the price elasticities of demand for imports and supply of non-mining exports, a new and lower equilibrium exchange is calculated, and the consequent resource cost and welfare changes as the economy moves to this rate are estimated. The loss to the economy which would result if the non-fuel mining sector were absent is equal approximately to one-third of the net foreign exchange (export revenue less directly imported inputs) generated by the mineral exports. This gain is additional to the RV figure calcualted by the team on the basis of the share of export revenue spent on locally produced products out of increases in mineral output or price (58).

The boost to the exchange rate from the growth of a large mineral exporting sector is the way in which additional imports are made available to the economy at lower real cost in terms of domestic resources. The higher exchange rate, however, damages the prospects of non-mineral exports and of import competing industries (though in the latter's case there is some offsetting benefit as demand for most commodities rises with the economy's increase in real income). If the exchange rate appreciation were the result of an expansion of agricultural or manufactured exports - whose existence might be expected to continue indefinately - the damage to other exports and to other forms of domestic economic activity might not be a matter of concern (59). Because mineral production is subject to exhaustion, there is likely to come a time when other exports will be needed to provide foreign exchange and when import competing industries also will need to expand; both requiring a depreciation of the exchange rate. Of course, in static reallocative terms this should not be a problem. In the context of long term development, however, there is a danger that domestic industries may be forced out of existence by imports while agricultural and other export activities may never develop to a stage where they can compete on world markets. More generally, a free foreign exchange market may fail to provide efficient intertemporal price signals and the exchange rate may settle at a level which would be approximate only if the present sales of mineral exports were to continue for ever. Also, a dominant minerals sector may not be able to provide sufficient employment, especially if a dualistic wage structure results from mining (60).

There has not been research time in this study to explore these effects on non-mineral sectors in Bolivia, but some general indications can be given. Bolivia certainly has not succeeded in becoming an exporter of manufactured products. In 1975 96 per cent of Bolivia's exports were of primary commodities (World Bank, 1980b). In 1963 agricultural products were only 4.6 per cent of total exports, the rest being minerals (principally tin and petroleum) though even this represented agricultural export growth compared with the previous decade (Zondag, 1965, p.177). Adverse effects on domestic agriculture via food imports have to some extent been counteracted by policies followed since the political upheaval of 1952 which led to a stress on agricultural development (especially by new settlement in the east and encouraged by transport improvements) and there was also a stress on petroleum exporting. Indeed, this chapter has shown that tin mining has been deprived of necessary reinvestment funds, in order to finance government expenditure elsewhere. Nankani's survey of mineral exporters notes that (like many other mineral export economies) Bolivia suffers from a relatively low savings ratio, but he stresses its high agricultural potential. Inflation has been high and that has been caused to a large extent by government deficit financing during slumps in mineral prices. On the other hand Bolivia, unlike many other mineral exporters, does not have any significant differential of mining wages relative to those in manufacturing, nor is the manufacturing wage:GNP per head ratio high compared with most LDC's. Thus at least Bolivia has escaped a dual wage structure (and in fact a later subsection will show that mining living standards are very low). Recent mineral supply problems have caused payments deficits and a currency devaluation, however (see subsection 3.3.2).

Since it is being argued that the tin industries in Malaysia, Thailand and Indonesia are 'marginal' to the economy in terms of their effect on the exchange rate, one would also not expect their effect on the general distribution of income in the economy to be significant. In any case in Malaysia (where tin has the largest share of exports, and exports have the largest share of GDP among the South East Asian producers) it will be shown later that mining wage differentials are not great in relation to manufacturing. In Thailand mining as a whole generates less than 2 per cent of GDP and tin in Indonesia less than 1 per cent. Also, in Thailand and (especially) in Indonesia tin mining is geographically distant from the main centres of economic activity and one would expect its influence to be diminished. Finally and more generally, it can be noted that in the more distant past in Malaysia tin was not a marginal activity and one of its effects (and in Thailand and Indonesia too) was to induce a large immigration of Chinese workers. But the subsequent role of the overseas Chinese in South East Asian development is certainly beyond the scope of this study (61).

5.4.2. Retained value

In chapter two retained value (RV) was suggested, with some reservations, as an indication of the distribution of gains between a host-country and foreign mining companies. RV shows that the extent to which a foreign owned activity has remained an 'enclave', and it is a convenient method of assembling data on local payments made out of export income, whose subsequent development effects can be traced. An increase in RV which results from a policy of integration backwards into intermediate products for the industry is subject to the proviso that the production should not be 'inappropriate' for local conditions. That issue is discussed in the next subsection, on linkages, though in fact many tin mining linkages have not developed as a result of government policy. Increases in RV resulting from increased taxes or from equity restructuring sometimes provide only short term gains, since they may damage foreigners' incentives to invest. This is considered in the next chapter in the context of investment appraisals, though it is worth noting at this stage that taxation levels, at least in South East Asia, do not seem so high as to destroy investment incentives. RV also is likely to vary between different production techniques, often associated in tin with different ownership. Thus encouragement to a locally owned sector such as gravel pumping or suction boat mining is sometimes an alternative or a supplement to restructuring the payments of a foreign owned sector to raise RV.

Tables 5.2, 5.4 and 5.6 show estimates of RV for Malaysia, Thailand and Indonesia based on cost breakdowns for 1978 supplied by the producer governments to the ITC. For Malaysia further very detailed splits of costs and payments are available from the Census of Mining Industries on a sectoral basis and from the Financial Survey of Limited Companies for tin mining as a whole; and these have been used to estimate the local content of each item. The latest available issue of the Census was for 1974 and of the Survey it was 1975 but the composition of individual cost terms such as power and materials is unlikely to have changed in the intervening three or four years. For Thailand and Indonesia similarly detailed further breakdowns are not available and more use has been made of general a priori information and rough estimates.

In Malaysia dredging RV is nearly 77 per cent, a rise from approximately 72 per cent in 1972. Wage and salary payments are a lower proportion of output value than in the early 1970's and the 1960's, when they averaged 11-15 per cent (62). The fact that they are about 90 per cent local reflects the fact that Malaysians hold all but a very few of the most senior positions. The most striking feature is the size of gross profits - nearly 66 per cent - reflecting the highly intra-marginal nature of Malaysian tin production and the resource rents generated thereby. Though the late 1970's was a time of exceptionally high tin prices, in 1972 the figure was still over 50 per cent and gross profits of nearly 70 per cent were also recorded in the 1960's. The increased local ownership resulting from Pernas' takeover of London Tin, discussed earlier in this chapter, results in a larger share of after tax profits remaining in Malaysia compared with the early 1970's, but to a large extent the high overall RV figure still derives from tax. Export duty has risen from about 16-17 per cent in the mid-1970's to nearly 29 per cent as tin prices have risen and the duty structure has been made more progressive. Income tax was first imposed on companies in 1948 at a rate of 20 per cent and was increased to 30 per cent in 1951 and to its present 40 per cent in 1959 (Lim Chong Yah, 1967, p.265). In addition a variable tin profits tax (63) has been levied since 1965 and a 5 per cent development tax since 1967. The export duty dates from the nineteenth century and in most post-war years it took about 15 per cent of output value.

Compared with the period before the Second World War, government policies, together with increased local production of inputs, have brought about a substantial change in the distribution of gains. In 1928, the first year for which sectoral output statistics are available, wages were less than 10 per cent of gross output value, export duty 14 per cent and other operating costs probably about 10 per cent. Given the then higher foreign content of wages and salaries, and the remission of profits overseas, RV could have been as little as 25-30 per cent.

As a check on the ITC cost breakdown Table 5.3 gives figures taken from the published accounts of the main Malaysia Mining Corporation companies for 1977-78. There is some difficulty in comparing the companies' financial year statistics with those of the 1978 calendar year in Table 5.2, but the profit figures are broadly similar, though probably indicating that the MMC companies have slightly higher profitability than the sectoral average. It should be noted though that since the 'other proceeds' involve investment income from cross-holdings there may be some overstatement of profits. Table 5.3 also provides a comparison with Thailand. The two Thai companies in the table, Aokam and Tongkah Harbour, are, as mentioned earlier in the study, part of the 'TASK' group, controlled in 1978 by MMC, and their output was over 90 per cent of that of the group (64). The broadly similar structure between the Thai and Malaysian operations is clear, though direct tax in Thailand is lower and export royalty higher. The figures also confirm the high share of profits in offshore dredging given in Table 5.4. In onshore dredging in Thailand profits are slightly lower, reflecting lower grades and older equipment. RV in offshore dredging in Thailand is lower than that in dredging in Malaysia. The difference is partly that the Thai operations use imported diesel as fuel instead of local electricity (65), but more because the table assumes that in 1978 local ownership was nil, prior to the late 1978 restructuring. Restructuring the equity to give 60 per cent Thai ownership would raise the RV figure almost to the Malaysian level. RV in onshore dredging is much higher in Thailand than for the offshore operations, partly because it is more labour-intensive and because it tends to consume local electricity, but also because it mainly is locally owned.

Comparing dredging in the two countries to gravel pumping indicates in Malaysia a higher RV for the local sector and in Thailand a higher RV compared with offshore dredging. The superiority of Thai onshore dredging to gravel pumping in terms of RV is largely because gravel pumping consumes imported diesel oil in large quantities while paying only a slightly higher share of its output value to local labour. Table 5.5 gives a breakdown of export revenue in the suction boat sector based on a small sample survey by a team from Chulalongkorn University. Taking into account considerable variation in the performance of individual boats, an overall impression of high profitability nevertheless can be obtained. Of the cost terms only the petroleum products, some of the repair components and a few items and expenditure for meals will have been imported, and it is difficult to see how RV could be much less than 90 per cent of current revenue, though probably much of the materials for the boats' constructuion will have been imported. Of course, RV will be lower for boats which smuggle their ore out of Thailand!

In all sectors in Malaysia and Thailand the high profitability before tax is clear. In Thailand, more reliance is placed on export royalty than in Malaysia, and like Malaysia this dates back to the last century. Taxes on company profits date from before the Second World War, but there has been a serious problem of non-compliance in the corporate sector generally (Ingram, 1971, pp.184-87, 299-301) so the reliance on a more easily collected output tax is understandable. High before tax profitability in dredging is also a feature of Indonesian tin mining (Table 5.6), especially offshore where grades are high. RV in Indonesian dredging of over 90 per cent reflects mainly the complete local ownership (in 1978) through P. T. Timah and also the use of local petroleum products. In gravel pumping profitability is surprisingly low, given the highly intra-marginal nature of almost all other South East Asian tin production but RV is high mainly as a result of the intensive use of labour and the use of local power. The high RV figures for Indonesian tin is confirmed by an estimate by P. T. Timah that the corporation's total imports in 1977 were $22.9 million, roughly 9 per cent of total export revenue (Interview, Jakarta, April 1979).

The generally high RV figures in South East Asia can be compared with the estimates for Bolivia for 1974 by the Harvard team headed by Gillis (1978, chapter three). Gillis was interested in the prospective multiplier effects of expansion in Bolivian mine production, and of increases in price (66), and the team estimated an RV of 45 per cent for an output increase at constant prices, and 52 per cent for a price increase alone. These estimates (which are for all non-fuel minerals and include private sector production, but are dominated by tin) are surprisingly low, given that most Bolivian production is controlled by Comibol. However, Gillis' definition of RV is different from that used here and from that of most other commodity studies in that its 'first round' refers to expenditure on locally produced goods and (non-factor) services rather than income accruing to domestic income recipients. In particular Gillis deducts from RV imports purchased by export sector workers and Bolivian owners of capital, and savings made by them, and also imports purchased by the government out of export tax revenue. Adding these items back into the RV estimates gives figures of 66 per cent (67) for the output change. It gives nearly 100 per cent for the price change, the benefits of which accrue almost entirely

TABLE 5.2

PERCENTAGE DISTRIBUTION OF EXPORT REVENUE, MALAYSIA, 1978

| | DREDGING | | GRAVEL PUMPING | |
|---|---|---|---|---|
| | TOTAL | LOCAL CONTENT | TOTAL | LOCAL CONTENT |
| COSTS | | | | |
| Power | 7.0 | 5.5 | 15.1 | 10.0 |
| Labour | 7.9 | 7.1 | 17.4 | 17.5 |
| Materials | 7.4 | 1.8 | 9.9 | 3.0 |
| Other Charges | 6.2 | - | 7.2 | 5.0 |
| Depreciation | 3.6 | 3.6 | 5.0 | 5.0 |
| Exploration and Development | 0.1 | 0.1 | 0.9 | 0.9 |
| Realisation Costs | 1.9 | 1.9 | 1.3 | 1.3 |
| TOTAL COSTS | (34.1) | - | (56.7) | - |
| ALLOCATION OF GROSS PROFITS | (65.9) | - | (43.3) | - |
| Export Duty | 28.6 | 28.6 | 31.1 | 31.1 |
| Tax | 17.0 | 16.0 | 6.0 | 6.0 |
| Dividends and Other Payments | 20.3 | 12.2 | 6.2 | 6.2 |
| TOTAL ALLOCATION OF EXPORT REVENUE | 100.0 | 76.8 | 100.0 | 86.0 |

Sources and Notes

(1) Breakdown of costs and export duty is from data supplied to ITC by producers, published in Engel and Allen (1979, pp.127 and 137). These are given on a unit cost basis, and unit export revenue is here assumed to be the average 1978 Penang price (M$1743 per picul). Since the average price is unweighted, some inaccuracy can result e.g. the export duty payment for gravel pumping is overstated slightly.

(2) The local content of each cost item is estimated on the basis of a priori knowledge, the more detailed cost breakdown in the Malaysian Census of Mining Industries (1974 being the latest available) and the 1970 Input-Output Tables. For example, in 1974 93 per cent of dredging's fuel purchases were of electricity, which in the 1970 IO table had a 20 per cent direct import content, which gives 5.2 percentage points local content. Assuming a two-thirds local content for other fuel purchases (mainly products/local petroleum refining), gives a total local content of 5.5 percentage points.

(3) Tax payments consist of 40 per cent income tax and 5 per cent development tax, rounded up to nearest whole number to adjust for tin profits tax payment which is difficult to calculate on a

Sources and notes (continued)

sectoral basis.

(4) Dredging's local dividend payments calculated on basis of 60 per cent local ownership (see chapter five).

(5) Totals may differ slightly from the sum of individual items because of rounding.

TABLE 5.3
PAYMENTS STRUCTURE OF MALAYSIA MINING CORPORATION COMPANIES IN MALAYSIA AND THAILAND, 1977-1978
(Percentages of annual value of tin output)

| | MALAYSIA | THAILAND |
|---|---|---|
| Operating expenses and overheads | 29.4 | 22.9 |
| Profit before tax and Malaysian export duty or Thai royalty | 71.3 | 72.3 |
| Net dividends | 19.5 | 14.6 |
| Tax to host-country | 23.9 | 20.1 |
| Export duty or royalty | 27.2 | 30.0 |
| Other proceeds | 7.8 | 5.5 |

Sources and Notes

(1) All data from company reports for 1977-78.

(2) Data for MMC companies operating in Malaysia exclude a small number of minor companies comprising 15-50 per cent of the output of the former London Tin Corporation. Among the companies operated by Charter Consolidated, Sungei Besi is excluded because (as a non-dredging company) its payments structure would differ from the others.

(3) Figures for operating costs and export duty are not given for the CC companies, and the table's figures therefore refer to LTC companies alone.

(4) Dividends are net of 40 per cent Malaysian income tax for the Malaysian operations and of United Kingdom income tax for the Thai operations. All companies are registered in Malaysia.

(5) The companies are (Malaysia) Berjuntai, Kamunting, Lower Perak, Malayan, Southern Malayan and Southern Kinta, and (Thailand) Aokam and Tongkah Harbour. Southern Kinta operated in both countries but is treated as Malaysian because two-thirds of its output comes from Malaysia.

TABLE 5.4
PERCENTAGE DISTRIBUTION OF EXPORT REVENUE, THAILAND, 1978

| | OFFSHORE DREDGING | | ONSHORE DREDGING | | GRAVEL PUMPING | |
|---|---|---|---|---|---|---|
| | TOTAL | LOCAL CONTENT | TOTAL | LOCAL CONTENT | TOTAL | LOCAL CONTENT |
| COSTS | | | | | | |
| Power | 4.2 | 0.3 | 9.7 | 7.5 | 15.6 | 1.0 |
| Labour | 7.2 | 6.2 | 11.6 | 11.0 | 12.2 | 12.2 |
| Materials | 6.9 | 1.7 | 8.1 | 2.0 | 8.3 | 2.0 |
| Other Charges | 6.9 | - | 1.77 | 1.5 | 2.1 | 2.1 |
| Depreciation | 3.3 | 3.3 | 2.5 | 2.5 | 3.0 | 3.0 |
| Exploration and Development | - | - | - | - | 1.0 | 1.0 |
| Realisation Costs | 2.7 | 2.7 | 3.1 | 3.1 | 3.3 | 3.3 |
| TOTAL COSTS | (31.3) | - | (36.9) | - | (45.5) | - |
| ALLOCATION OF GROSS PROFITS | (68.7) | - | (63.1) | - | (54.5) | - |
| Export Royalty | 29.5 | 29.5 | 29.4 | 29.4 | 30.2 | 30.2 |
| Tax | 16.1 | 15.1 | 16.1 | 16.1 | 12.8 | 12.8 |
| Dividends and Other Payments | 27.4 | - | 17.6 | 13.2 | 11.5 | 11.5 |
| TOTAL ALLOCATION OF EXPORT REVENUE | 100.0 | 58.8 | 100.0 | 86.3 | 100.0 | 79.1 |

Sources and Notes

(1) Cost and Royalty breakdown adapted from Engel and Allen (1979, pp.127 and 137) (see notes to Table 5.2). Local content estimated on basis of *a priori* information. For onshore dredging, Fairmont State is the only company with foreign connections, though apparently majority Thai owned. According to *TI* August 1979 its 1978 output was approximately half of the sector's total, so if 49 per cent foreign ownership is assumed, then a quarter of the sector's profits may be available for remission overseas, assuming profitability differs little between companies.

(2) Company tax is 35 per cent (30 per cent on companies listed on Thai stock exchange) and there is a 4.4 per cent business tax on gross proceeds less realisation charges.

TABLE 5.5
PERCENTAGE ALLOCATION OF EXPORT REVENUE
IN SUCTION BOAT MINING, THAILAND, 1976

| | |
|---|---|
| COSTS | |
| Diesel, benzene, engine oil | 7.0 |
| Repairs | 3.3 |
| Wages to hired labour | 2.0 |
| Meals | 5.9 |
| Interest charges | 4.4 |
| Workers' share of revenue | 22.7 |
| ALLOCATION OF GROSS PROFIT | (54.7) |
| Owners' share | 18.3 |
| Export royalty | 20.9 |
| Income tax | 0.2 |
| Provincial administration service charge | 1.6 |
| Authorised dealers' charges | 13.6 |

Sources and Notes

(1) Calculated from Chulalongkorn University (1976, pp.25 and 30).

(2) Refers to boats operating in Phanggna province.

TABLE 5.6
PERCENTAGE DISTRIBUTION OF EXPORT REVENUE, INDONESIA, 1978

| | OFFSHORE DREDGING | | ONSHORE DREDGING | | GRAVEL PUMPING | |
|---|---|---|---|---|---|---|
| | TOTAL | LOCAL CONTENT | TOTAL | LOCAL CONTENT | TOTAL | LOCAL CONTENT |
| COSTS | | | | | | |
| Power | 3.9 | 2.6 | 6.5 | 5.2 | 10.5 | 7.0 |
| Labour | 8.5 | 8.5 | 15.4 | 15.4 | 27.4 | 27.4 |
| Materials | 7.1 | 1.8 | 10.1 | 2.5 | 16.6 | 4.1 |
| Other charges | 10.4 | 10.4 | 14.3 | 14.3 | 22.2 | 22.2 |
| Depreciation | 1.4 | 1.4 | 2.4 | 2.4 | 4.0 | 4.0 |
| Exploration and Development | 1.3 | 1.3 | 1.6 | 1.6 | 3.7 | 3.7 |
| Realisation Charges | 5.4 | 5.4 | 5.2 | 5.2 | 5.2 | 5.2 |
| TOTAL COSTS | (38.1) | - | (55.7) | - | (89.7) | - |
| ALLOCATION OF GROSS PROFITS | (61.9) | - | (44.3) | - | (10.3) | - |
| Export Tax | 9.7 | 9.7 | 9.8 | 9.8 | 9.7 | 9.7 |
| Company Tax | 23.5 | 23.5 | 15.5 | 15.5 | 0.3 | 0.3 |
| Other Payments | 28.7 | 28.7 | 19.0 | 19.0 | 0.3 | 0.2 |
| TOTAL ALLOCATION OF EXPORT REVENUE | 100.0 | 93.3 | 100.0 | 90.9 | 100.0 | 83.8 |

Sources and Notes

(1) Cost breakdown from Engel and Allen (1979, pp.127 and 137).

(2) Corporation tax is levied at 45 per cent.

TABLE 5.7
PERCENTAGE DISTRIBUTION OF EXPORT REVENUE IN UNDERGROUND MINING IN BOLIVIA AND AUSTRALIA, 1978

| | BOLIVIA | AUSTRALIA |
|---|---|---|
| COSTS | | |
| Power | 2.5 | 2.1 |
| Labour | 25.6 | 15.4 |
| Materials | 12.1 | 15.1 |
| Other charges | 9.5 | 5.3 |
| Depreciation | 1.9 | 6.5 |
| Exploration and Development | 3.3 | 2.3 |
| Realisation Charges | 23.1 | 2.0 |
| TOTAL COSTS | (77.9) | (48.4) |
| ALLOCATION OF GROSS PROFITS | (22.1) | (51.2) |
| Export Royalty | 23.7 | 1.5 |
| Corporation Tax | - | 22.9 |
| Other Payments | -1.6 | 26.8 |
| TOTAL ALLOCATION OF EXPORT REVENUE | 100.0 | 100.0 |

Sources and Notes

(1) Cost structure and Australian tax rate from Engel and Allen (1979, pp.127, 137 and 156).

(2) Although the table refers to 'export revenue', note that a substantial part of Australian tin production is consumed locally.

TABLE 5.8
ECONOMIC COSTS AND BENEFITS AS PERCENTAGES OF
TOTAL VALUE OF OUTPUT IN SOUTH EAST ASIAN TIN MINING

| | ECONOMIC VALUE OF OPERATING COSTS | DEPRECIATION AND INTEREST CHARGES | NET ECONOMIC GAIN | NET ECONOMIC GAIN AFTER PROFIT OUTFLOW |
|---|---|---|---|---|
| MALAYSIA | | | | |
| Dredging | 24.3 | 15.0 | 60.7 | 52.6 |
| Gravel Pumping (Large Mines) | 32.3 | 6.7 | 61.0 | 61.0 |
| Gravel Pumping (Small Mines) | 49.2 | 9.1 | 41.7 | 41.7 |
| THAILAND | | | | |
| Onshore Dredges | 42.6 | 22.8 | 34.6 | 30.2 |
| Gravel Pumping | 38.1 | 10.9 | 51.0 | 51.0 |
| INDONESIA | | | | |
| Offshore Dredges | 22.7 | 20.6 | 56.7 | 56.7 |
| Gravel Pumps | 78.1 | 6.0 | 15.9 | 15.9 |

Sources and Notes

(1) All figures derive from the investment appraisals of chapter six, with output values and smelting charges adjusted to refer to the average annual 1978 tin price (M$1743).

(2) The projects used here are (in order) DA, GPA and GPC for Malaysia; DSH and GPC for Thailand; and DB and GPC for Indonesia. See Table 6.1 for project details.

(3) Interest calculated at 10 per cent p.a. on the declining balance and all capital expenditure is treated as if it occurs the first year of pre-production.

to the government, local workers and local owners of capital. The fact that the Gillis estimates are for marginal propensities whereas the figures here are averages for the whole of a sector is not in practice an important difference, since almost all of the Bolivian marginal estimates assume the average and the marginal values are the same and are derived from statistics for averages.

In view of the detailed nature of the Bolivian estimates and the fact that 1974, like 1978, was a year of high tin prices, it did not seem worth restructuring them on the basis of 1978 data. Instead the 1978 ITC cost breakdown for Bolivia is given in Table 5.7, and Australian statistics are added for comparison. The much lower profitability of Bolivian mining compared both with Australian underground operations and with alluvial mining in South East Asia is apparent from the table, and the high Bolivian smelting charge (and other components of the realisation charges) is striking. The heavy export royalty in fact generates an accounting loss in Bolivia.

In chapter two RV was presented as a method of measuring the net foreign exchange contribution and the real income generated by (marginal) mineral export activities. The net foreign exchange contribution, which can be inferred directly from the RV figures set out above, is relevant principally where the country in question suffers from a chronic balance of payments deficit which it cannot cure by traditional methods. All three South East Asian producers in the late 1970's had reasonably strong balance of payments positions, though Thailand's could well worsen if its agricultural productivity does not increase and Indonesia may face problems as a result of growing rice imports, especially if its oil production stagnates. Bolivia's position is very much less comfortable, with an external debt service ratio equal to 49 per cent of its export earnings and a current account deficit equivalent to 48 per cent of the value of its merchandise exports (World Bank, 1980b, pp.124 and 134); but its 1979 devaluation may help (see subsection 3.3.2).

To assess the real income contribution of an export activity involves evaluating the real resource costs of the components of RV. In Gillis' (1978, pp.68-69) account of Bolivia, the RV figure for changes in output is assumed equal to the real income gain by several simple devices. First, it is assumed that the labour used has no opportunity cost, and second, it increases the assumed cost of petroleum to take into account the fact that it sold domestically at below world market price. Inputs are mostly imported and those are taken at their foreign exchange cost. This, as noted, gives a figure of roughly 66 per cent when adjusted to compare with the South East Asian estimates; though for the Bolivian mineral export sector as a whole (see subsection 5.4.1) the real income gain is equivalent to a third of net export revenue when ramifications elsewhere in the economy are considered. For South East Asia, the project appraisals of chapter six provide data which allows a more detailed measure of real income gain to be estimated, analagous to the Lall-Streeten methods for import-substituting projects discussed in chapter two. Here all inputs (including labour) are revalued at efficiency prices (68) and capital costs (also valued at efficiency prices) are charged at an assumed 10 per cent rate of interest. These costs are deducted from the value of output at the 1978 average tin price and the difference is expressed as a percentage of output value, yielding a net economic gain coefficient. These are shown in Table 5.8. These coefficients, since they are drawn from individual projects, may not be as typical of their sectors as are the ITC figures on which Tables 5.2, 5.4 and 5.6 are based (69). The cost figures in the two sets of data are not strictly comparable since in Table 5.8 all assumed interest charges are treated as a cost, whereas in the others some may be included in dividends where the investment was financed by equity. Generally the costs in financial terms are higher than those in economic efficiency prices (except for Indonesian petroleum sold to tin mining at below the world price). For Malaysian dredging, according to Table 5.2, about 8 per cent points should be deducted from the gain coefficient to take account of the outflow of dividends to abroad, and for gravel pumping note that the small mine is still the more common sort to Malaysia.

In general Table 5.8 illustrates the high economic efficiency of tin mining in Asia, especially in Malaysia, and in Indonesia mainly in dredging.

The coefficients include local payments of export duty/royalty and tax. It is worth asking whether the high rates of export duty charged in three of the main producers (Bolivia, Thailand and Malaysia) have increased local gains by shifting the incidence of the tax on to foreign consumers and increasing revenue in the face of inelastic demand. A study by Lam (1978) of tax incidence in Malaysian tin argues against this proposition, on the grounds that Malaysian miners face a virtually infinitely elastic demand and that they have a very low supply elasticity. In fact, the imposition of a heavy export royalty by Bolivia since 1965 (70) may well have curbed production by depriving Comibol of funds for replacement investment and marginal (gravel pump) mines in South East Asia are adversely affected (as chapter six will show) by heavy taxation on output. Moreover, as more recent estimates (by Bird, 1978) have become available, there are grounds for believing that the estimates of tin supply elasticity in the studies cited by Lam are too low. Thus heavy export taxation may have reduced supply and may have been shifted in part to consumers (71). The very high prices of the late 1970's provide some support for this possibility, though the tax level and structure have been by no means the only adverse influence on supply during that period. Regarding demand elasticity, it is doubtful that a large production cutback by a major producer would leave price unchanged.

### 5.4.3 Backward linkages

The large mineral rents generated by tin in South East Asia have made tax and ownership policy an important part of securing development gains. Scope has also existed for backward linkage creation, though much of this has come about naturally through market forces. As the tables in the previous subsection have shown, power usually accounts for between nearly 5 per cent and over 15 per cent of gross output value. Gravel pumps and offshore dredges normally use diesel and onshore dredges use electricity. In Indonesia electricity and diesel, in roughly equal proportions, accounted for 60 per cent of total intermediate purchases by the tin ore mining sector in 1971 (Tabel Input-Output, Indonesia, 1971), and this understates the importance of diesel which is sold in Indonesia below the world price. In Malaysia fuels constituted 41 per cent of dredging's purchases of materials and 51 per cent of those of gravel pumping in 1974 (Census of Mining Industries, Peninsular Malaysia

1974) (72), Indonesia and Malaysia are net exporters of petroleum products. In 1971 mining purchases constituted about 25 per cent of the output of the Malaysian petroleum products industry. The existence of this large consumer may well have influenced the Malaysian government in its decision to give 'pioneer' (tax exempting) status to Esso and Shell to establish local refineries in the 1960's, and the available evidence suggests that petroleum refining in Malaysia is not a 'bad' linkage (73). Electricity generation on a large scale dates from 1928 in Malaysia, and in the late 1930's tin mining bought over 30 per cent of the units generated in the two main west coast states of Perak and Selangor. In Indonesia the mining islands have their own electricity generating facilities though, since the islands are far from the country's main centres of population, the facilities do not yield externalities to other industrial users. In Thailand mines in the main tin area in the south purchase electricity on a large scale but the development of the industry was not investigated (74).

Railway development in Malaysia also depended heavily on the tin industry, in the sense that the railways were built largely over the 1884 -1910 period out of government revenue, over a third of which came from the export duty on tin - a good example of 'fiscal linkage' (Hirschman, 1977, pp.72-77). In colonial Malaya the early stages of railway development had taken the form of lines from the tin areas to the ports, and the national network was formed largely by joining these up. In Thailand although railways had been built by the government since the late nineteenth century and a line has linked the South (the main mining area) with the central region since before the Second World War, one cannot link railway development to tin in a similar fashion. Tin never occupied the dominant position as a source of revenue that it did in Malaya in the early years of the industry's rapid development, and tin ore was shipped directly out to Malaya for smelting. In Indonesia tin was shipped abroad directly from the mining islands.

In Malaysia probably the most interesting linkage, though it owes little to government action, is the growth of a local light engineering industry to meet the needs of the tin industry both for spare parts and for capital equipment. Engineering parts are the most important intermediate product purchased by the tin industry after fuel, and local engineering firms date to before the First World War. With regard to capital equipment, about a quarter of the value of an inland tin dredge and about 40 per cent of the capital cost of a gravel pump operation consist of locally purchased engineering and construction items. In Indonesia, extensive facilities exist for the manufacture and servicing of mining equipment at P. T. Timah's workshops on Bangka, which are capable of doing dredge rehabilitation and which date from Dutch colonial times. They do not extend to major new construction, however, and two offshore dredges commissioned in recent years (Banka I in 1966 and Banka 2 in 1978) were built overseas (in the UK and Japan) and towed to site. This is also true of offshore dredging in Thailand, though in Malaysia dredges are normally fabricated on site. In Thailand local workshops in the Phuket area produce a range of spare parts (aided by a 15 per cent import duty). In Malaysia and Thailand most of the engineering industry is in the hands of locally owned (and mainly ethnic Chinese) firms, though dredge design is normally done overseas. Malaysia dredge erection contractors have also carried out work in Thailand and Indonesia.

In alluvial mining backward linkage generation is qualitatively similar between the various alternative techniques, though (as already noted) dredging tends to use electricity as fuel and gravel pumping uses petroleum products. As the tables earlier in the section show, gravel pumping tends to be more materials intensive. In Thailand this leads to a larger outflow of export revenue than in inland dredging, since Thailand is not a petroleum producer, but it also leads to a larger demand by the sector in all three countries for engineering parts. The demand for engineering products has also been greatly enhanced in Thailand by the growth of suction boat mining, the boats being of local construction. A detailed case study of engineering in Malaysia (in Thoburn, 1977b, chapter eight) showed little evidence of its being a 'bad' linkage. Indeed it has proved a useful vehicle for the development of small and labour-intensive businesses. Further scope exists for local manufacture of mining supplies, such as an increased local output of diesel engines in all three producers, but careful project appraisal will be necessary if these are to be made as part of an import substitution policy behind tariff barriers.

Overall in Malaysia, according to the 1970 input-output table, only a quarter of the mining industry's purchases of intermediate products (which are overwhelmingly for tin mining) was imported directly (75). In Bolivia in contrast, Gillis' team (1978, p.76) estimated that 85 per cent of the material purchases of non-fuel mining were imported. It will be observed from Table 5.7 that (local) power is a small item in Bolivia's mining costs (though its importance is understated to the extent that Bolivian petroleum products are sold to the industry at less than the world price), and it appears therefore that virtually everything else is imported.

Finally, it is important that the backward linkages generated by tin, at least in Asia, are such as to outlive the eventual exhaustion of tin production. The provision of power relies increasingly for its market on other industrial development, and engineering firms can find a wide range of customers outside of the mining sector (76).

5.4.4. Forward linkages

Forward linkages involve moving from a consideration of RV to looking at the producer countries' share of Final Consumer Value.

The local development of tin smelting has already been traced in some detail and all South East Asian tin output is now smelted in the region. Smelting adds only about 1-5 per cent of output value alluvial mining but it saves 25 per cent weight in shipping and improves the countries' bargaining position relative to purchasers. The Malaysian smelters, at least, have been extremely profitable historically. It was estimated in 1974 that the Straits Trading Company made a M$12 million profit on a probable net turnover of M$M18 ($M749 million gross turnover less payment of M$733 million for supplies of unsmelted tin). High past profits have been put into investments, including some mining companies. Straits Trading derives almost half its profits from investment (AR 1977). In 1969 Datuk Keramat owned investments of M$31 million compared with its own total market value of M$75 million (Thoburn, 1977b, pp.123-27). Smelting charges in Thailand are geared to those of the Malaysian smelters but data on Thaisarco's profitability are not available, nor on that of Peltim in Indonesia. However the processes used (after some initial experimentation in Indonesia, already discussed) are basically similar to those in Malaysia, and there is no reason to believe that the Malaysian smelters' high private profitability greatly overstates their social profitability since their output is exported on a free market and their inputs are not subsidised. The problems faced by Bolivia of establishing domestic smelting capacity for her complex, hard rock ores also have already been discussed and there may well be some trade off between improving the country's bargaining position against foreign buyers and the high real costs of domestic smelting.

Forward integration into the manufacture of tinplate and tincans is the other main potential linkage, and there is a local demand in South East Asia not only by consumers but from export industries selling canned fruit such as pineapples. The problem here is that tinplating is more closely allied with the local establishment of steel making than with tin mining (since tin metal is a tiny part of total cost) and tinplate for can manufacture can be imported, so neither activity is greatly aided in cost terms by the existence of local tin mining. Thus the anti-linkage argument of Little-Mirrlees, discussed in chapter two, clearly applies here. It contrasts to the backward linkages of electricity generation, where lack of transportability gives natural protection to domestic capacity, and engineering where local contact with customers is extremely important (77). A recent study for UNIDO by David Lim (1980) was extremely pessimistic about tinplate manufacture, and gave instructive data about the tinplate industry in Thailand, the only major tin producer to have one. The Thai industry consists of one firm, which is a joint venture between local interests and a Japanese steel manufacturer, and runs at under a third capacity while more than half of Thai tinplate consumption is imported. The main problem is that much of Thai demand is for lower quality ('secondary') tinplate whereas the local company produces 'prime' tinplate. However, nor can it export to neighbouring Malaysia and Indonesia, because of high costs and their preference for 'secondary' tinplate, and Lim does not feel it has prospects of exporting to industrial countries. In any case, even if all tinplate consumed in Thailand (including that of local pineapple canning) were produced locally it would only require about 3 per cent of the Thai production of tin metal. In Indonesia domestic demand is smaller still and below the efficient minimum production for an electrolytic plant. Tinplate has a very high income elasticity of demand (Lim, 1980, p.210) so LDC tin producers appear to need to wait until their domestic markets grow. However, Brazil's tinplating industry (the only other LDC tin producer at present with one) appears to work at near capacity, and is reported to be expanding (TI, January 1980). Malaysia's demand for tinplate has grown rapidly too, apparently (at 10-20 per cent per annum since the early 1970's) (78), and it was recently announced that a major tinplate plant would be established as a joint venture between the government owned food corporation FIMA and Japanese steel interests with a minority holding (35 per cent) taken up by MMC, the Johore State Economic Development Corporation and some pineapple canners (TI, November 1979). In view of Thailand's experience the social gains from this venture must remain provisional. Indonesia is also said to have plans to introduce tinplating in a steel complex in Java, but Lim cites a 1972 feasibility study by P. T. Timah which reports strongly against local tinplating. Bolivia at present has no tinplate facilities.

Besides tinplate, solder appears to have some prospects, especially in Malaysia where there is a flourishing electronics industry. Tin, though constituting 60 per cent of the weight of the solder, is only a minute component in the total cost of the final electronics product so, again, the proportion of local metal output which would be taken would be very small.

5.4.5 Final demand linkage

In South East Asia data on local market creation by tin mining workers are available only for Malaysia, where it is estimated that roughly 70 per cent of their expenditure is on locally produced goods (Thoburn, 1977b, p.179). In Bolivia the Harvard team estimated that 75 per cent of mine workers' expenditure was local. These high local proportions also could be true for Thailand given the ethnic similarity of the workforce to that of Malaysia, though obviously consumption would be influenced too by the pattern of protection. In all three Asian producers the absence of large numbers of really highly paid workers found in some mining industries elsewhere again predisposes one to expect a low import content in expenditure.

Urbanisation is also an important feature of local market creation (79), and in Malaysia urban growth was associated with tin mining. Most of the present day major cities - Kuala Lumpur, Ipoh, Seremban - originally were mining towns. In Thailand and Indonesia in contrast tin mining areas were far from other main centres of population. In Bolivia the mountainous region has both the bulk of the country's population and almost all its mining, but this area is far removed from the agricultural areas of the country and difficult to reach from them. To that extent it has not led to the economic integration of the country. The transport system has been geared to shipping ore out via Chile, whereas in Malaysia the rail network built to transport tin has formed the nucleus of the present day transport system, and its role in facilitating the spread of rubber cultivation and other economic activity is well documented.

With regard to the effects of tin expansion on local demand, there is no good reason to suppose that expanding the workforce marginally will significantly change the pattern of consumption with regard to import content.

5.4.6. Employment creation and unpriced externalities

Although the tin industry's past record of employment intensity is of some interest, the attention of policy makers naturally centres on the possibilities of generating new employment by investment in new projects. In consequence much of the discussion is best left until tin investment appraisals in the next chapter where employment is considered in section 6.6. Reference forward to that section (and to Tables 6.12 and 6.13) will show, both for the investment projects considered and for the sectors as a whole, that gravel pumping in Asia has a much higher employment intensity generally than dredging, especially dredging offshore (though the onshore dredging project in Thailand is unusually labour-intensive). It will be remembered that both Comibol in Bolivia and Timah in Indonesia run their operation at a somewhat higher labour intensity than commercial profitability would suggest. With regard to investment, the very high employment creating potential of gravel pumping (Table 6.12) is striking, though one should remember that its capital equipment is shorter lived than that of a dredging project. Chapter six will also show that the employment creating effects of suction boat mining many times exceed that of offshore dredging, both in output terms and investment terms, though again project life is shorter.

Of the unpriced externalities generated by tin, labour training is probably the most important. About 40 per cent of the workforce in a Malaysian dredging operation is skilled or semi-skilled. In gravel pumping the figure is nearly 30 per cent (Thoburn, 1977b, p.213), though the higher overall labour intensity of gravel pumping would mean that the overall use of skilled workers per unit of output and per unit of investment is greater by the gravel pumping sector. Clearly in the short run the increased use of skilled workers as a result of an output expansion in tin would be a cost to the economy, but in the longer run a higher demand for skills tends to be met by an increased supply and an overall improvement in the quality of the workforce. Moreover the skills used in tin mining are usually usable by other sectors if the workers wish to move - drivers, electrical chargemen, skilled workshop workers, for example, are prominent in the skilled mining workforce. The skilled nature of the mining workforce is reflected in relatively high rates of pay in Malaysia in the sense that gravel pumping workers receive slightly higher wages than manufacturing workers on average and dredging workers' wages are 50 per cent higher. However, the employment intensity in tin mining in Malaysia is roughly equal in dredging and manufacturing, while gravel pumping's employment intensity is higher, so moderately high wages are not accompanied by a particularly low level of labour use (80).

In Thailand and Indonesia there is no reason to suppose a radically different skill mix in each of the sectors from that of Malaysia. Some evidence collected for chapter six (see the Annex to that chapter) suggests that the overall mining wage differential relative to shadow wages in the economy may be greater in Thailand than Malaysia. Casual empiricism suggests that this will also be true in Indonesia (shadow wage data for Indonesia in chapter six were very rough and ready at the time of writing).

In Malaysia there have been no Europeans in the artisan, junior supervisory or clerical positions since the very early days of the industry and the use of Thai nationals to replace expatriates dates from the Second World War. Even dredgemasters in Malaysia and Thailand are now often local, and in Indonesia local people in P. T. Timah have carried out all functions since nationalisation, as in Bolivia.

Data on the skill mix in Bolivia tin mining are not readily available, but the low living standards of workers in Comibol and other Bolivian mining companies have been noted by a variety of commentators. Most recently a British trades union delegation (81) remarked on the 'appalling wages' of Bolivian mine workers and noted that miners' life expectancy was on average only 35 years while silicosis and tuberculosis were endemic and mine safety precautions were almost non-existent (82). Thus the low accounting profitability of Comibol, the high costs caused by the nature of the tin deposits and the lack of reinvestment in the post-war period have resulted in workers enjoying lower real incomes than most tin mine workers in Asia. The British delegation quotes 1979 wages as £0.80 per day minimum, though noting that some workers can earn more by bonuses and that they receive certain subsidies in foodstuffs at company stores. In 1976 unskilled workers on Malaysian tin dredges earned on average M$8.2 per day, roughly double the Bolivian rate, and cheap housing supplied to dredging workers is undoubtedly of higher quality than in Bolivia. In 1974 the average wages and salaries bill per employee per day on gravel pump mines in Malaysia was M$7.8 (83), and inflation has raised wages since then. In gravel pump mines visited in Malaysia in 1978 average wage bills were M$12 (c.£2.50) per worker per day plus another M$4 in free food. Similarly on a locally owned dredge in Thailand in 1978 the total wage bill was roughly £2.50 (100 bahts) per worker per day including overtime, though in a gravel pump mine in northern Thailand daily paid workers received only 30 bahts (plus free housing and medical attention). In Indonesia in 1979 an unskilled worker in Belitung earned £1.30 per day plus (often) free housing and rice, and on Bangka island about half as much again.

NOTES

(1) These details are from Zondag (1966). Geddes (1972) is useful on the background to nationalisation.
(2) This point is made by Geddes (1972, p.258), Zondag (1966, p.87) and Gillis (1978, p.28).
(3) Note though that this system was not a creation of the MNR government. It had been used to tax the mining sector since the 1930's (Geddes, 1972, chapter twenty-two).
(4) Similarly Zondag (1966, p.88) estimates that some 200 foreign technicians left Bolivia after nationalisation.
(5) This turnover figure includes purchases from private mines and cooperatives. In the mid-1970's these purchases accounted for about a quarter of Comibol's sales (Gillis, 1978, p.31).
(6) The reverse is also possible, though rarely in export orientated projects.
(7) The 1971-77 figures are from a study by D. J. Fox quoted in detail in TI, December 1978. More detailed figures for 1973-74 are given in Gillis (1978, pp.329-331) and for 1978 in Engel and Allen (1979, p.127).
(8) More details of the payments structure are given in the last section of this chapter.
(9) This is less than the amount spent by a number of mining companies in Indonesia on single claims (Gillis, 1978, p.50).
(10) These figures are understated to the extent that Comibol's sales include tin purchased from the private sector.
(11) Unless otherwise stated, the details about medium and small mines are from this source.
(12) According to Engel and Allen (1979, p.153) foreign ownership in private sector tin production in Bolivia is limited to 49 per cent.
(13) As reported in TI, August 1979.
(14) Calculated from P. N. Timah (1972) and excluding the negligible capacity figures for dismountable and pioneering dredges.
(15) Interview, P. T. Timah, Jakarta, April 1979. Unless otherwise stated, other information on Timah's current activities also comes from interviews in April 1979.

(16) Prior to 1976 Timah paid 25 per cent of its after tax profits to the government (as a contribution to the National Development Budget). In 1979 Timah's dividend payment was fixed at 20 per cent.
(17) Badan Koordinasi Penanaman Modal.
(18) In 1977 plans for Rp.14 billion (US $34) were submitted but only Rp.11 billion ($2 billion) were allowed to be counted against tax. There is provision for companies rehabilitating their assets to get an allowance of 20 per cent of the total cost, which can be counted against tax at the rate of 5 per cent a year.
(19) Based on data in P. N. Timah (1972) and P. T. Timah (1976).
(20) See chapter six for further discussion of the relative efficiency of various producers. Table 6.13 gives labour productivity figures based on mining labour statistics more narrowly defined and on a sectoral basis. Table 6.4 gives data on costs per unit of throughput for various projects.
(21) Outputs from Department of Mines and Energy, Indonesia (1977) and costs from P. T. Timah, Annual Financial Report, 1977.
(22) In 1973-74 Comibol produced 66 per cent of Bolivia's lead production, 88 per cent of zinc and 91 per cent of bismuth (Gillis, 1978, p.32).
(23) Certainly some workers on contractor's mines visited in April 1979 were Chinese.
(24) The 1969 Annual Report of P. N. Timah lists unsuccessful negotiations with Charter Consolidated and a consortium of Rio Tinto Zinc and Bethlehem Steel, all of whom have connections with tin in Malaysia or Thailand.
(25) Information from interviews with P. T. Timah and the three companies, April 1979.
(26) See Gillis (1978, pp.134-35) for a general discussion of Indonesian mining contracts. Only one 'first generation' contract was signed.
(27) Sixteen 'second generation' contracts were signed (US Embassy Indonesia 1977).
(28) The difference between the royalty and the export tax is that the former is levied on the assumed metallic content of tin concentrates delivered to the smelter and the latter on the value of the metal produced (Gillis, 1978, pp.135-38). Difficulties arise if the assumed content of the concentrates differs from their actual tin content, and BHP recently negotiated a change in assumed content.
(29) Only one such has been signed - with Rio Tinto Zinc on a copper project - and it included a 60 per cent windfall profits tax when the return on capital rose above 15 per cent over a three year period, together with the obligation to offer 51 per cent of tin equity (instead of the usual 25 per cent) within ten years of starting production. Contracts henceforth will have still tougher provision including stipulations about the use of local contractors (FEER, 1 February 1980). Though these provisions seem close to what many economists would recommend (see chapter two) they appear to be discouraging foreign investment in minerals.
(30) See Thoburn (1977b, p.127) on the Straits Trading Company. In 1978 the large trading company Amalgamated Metal Corporation which owns the other Malaysian smelter, Datuk Keramat, stated in its Annual Report that tin smelting accounted for 74 per cent of its profit while generating only 23 per cent of turnover. The figures for 1977 were similar.
(31) For Thailand no statistics are available to compare with Malaysia but the Billiton company which owns the Thaisarco smelter claims to make the bulk of its profits in the country from smelting rather than marketing. See next section.

(32) Because a large part of Indonesian output comes from offshore dredges, total output is lower during the stormy season, and a larger smelting capacity is necessary to deal with this seasonal pattern than the simple figure for total annual output would suggest.
(33) In 1978 the Straits Trading Company charged approximately M$15 per picul of ore, including transport and the energy surcharge (Interview, September 1978). Timah in 1979 quoted 1978 costs per tonne ore as Rp.108000. This is equivalent to M$38 per picul at the 1978 exchange rate and MS26 at the 1979 rate. The Indonesian cost would probably include depreciation whereas the STC's capital costs will have been long since written off. Also, the Mentok smelter regards itself as having social obligations to employ a larger workforce than necessary on commercial grounds and in the light of Indonesia's labour surplus this must be counted as a benefit.
(34) Widyono (1977, p.24) notes that from 1953-61 the Williams Harvey smelter, owned by the Patiño organisation, deducted nearly $20 million (in the form of a 10 per cent discount) as an 'advance' on indemnification payments for the nationalisation of the 'big three' mines.
(35) Charles Raw's (1978) 'exposé' of Slater Walker unfortunately has little to say about its Asian activities.
(36) According to Pernas' chairman Tengku Razaleigh (Tin, April 1980) the first discussions with Slater Walker were held in December 1972. Slater Walker's subsidiary Haw Par bought a 29.2 per cent stake in LTC in May 1973 but Pernas waited until April 1975 to make its purchase because it thought the shares were too expensive.
(37) For further details of the takeover and subsequent events, and on which this account relies heavily, see FEER, 1 April 1977, 7 September 1979 and 30 November 1979.
(38) See previous chapter.
(39) There was insufficient research time to permit an analysis of the share registers, company by company.
(40) Berjuntai, the largest company, is said to be 74 per cent locally owned (Interview, Malaysia Mining Corporation, September 1978).
(41) This is a weighted average figure derived from share details in Zorn and Leigh-Hunt (1974).
(42) Information from interviews in November 1978.
(43) See text of speech by Minister of Industry on Mineral Development Policy, 22-23 March 1978, translated from Thai by the Department of Mineral Resources, Bangkok.
(44) See Chulalongkorn University (1976, pp.29-30) for further details.
(45) Many of these complaints were aired at a national seminar on the mining industry in August 1977. See Thoburn (1978b).
(46) Speech by chief minister of Selangor, July 1978, printed in Tin, November 1978.
(47) As of 1979 38 per cent of Pacific's equity was owned by Denison Mines of Canada (TI, April 1979) though Pacific is often referred to as being 'US controlled' (e.g. TI, July 1979). Pacific owns some minor interests in Thai gravel pumping.
(48) Details of projects announced in 1979 and 1980 are from various issues of TI.
(49) This information is from a Thai Ministry of Commerce report (Sriwichchar and Pithaya, 1977).
(50) See TI, November, 1978. This issue also carried a report that a small mining subsidiary of Thai Present was owned by Thai, Belgian and Israeli interests.

(51) The exchange will be serviced by the International Commodities Clearing House of Sydney and a consortium of banks will hold 70 per cent of the equity (TI, March and May 1980). For some background information on the exchange see Murray (1977).
(52) Details from interviews in Brussels, June 1978.
(53) These details from Engel and Allen (1979, chapter four).
(54) Whereas RV and the local share of FCV are in themselves exclusively marginal measures, the measures of some 'wider' development effects merge into the non-marginal. In particular, the income distributional effects of the mineral export activity clearly should be considered on a non-marginal basis.
(55) Backward and forward linkages, of course, are analysed as part of RV and FCV.
(56) See also the short summary by Nankani (1980).
(57) As noted in section 2.2 this rise would have to be in relation to non-traded goods since the international terms of trade are held constant by assumption; though the Harvard analysis does not make the role of NTG's explicit.
(58) As the next subsection will show, the Gillis definition of RV is narrower than that generally used.
(59) Though a high degree of specialisation leaves the economy vulnerable to shifts in export demand, and the export industries would have to create sufficient employment.
(60) These comments on the exchange rate and the wage structure rely heavily on Nankani (1979).
(61) In general there is a problem in estimating the non-marginal effects of an export activity which changes the social structure of an economy considerably. For example it would seem slightly unreal to describe the 'without-rubber' situation in Malaysia, where massive importation of Indian (and Chinese) workers occurred and where the economy of Malay village society was radically altered in much of the country.
(62) These details of RV and the input structure in earlier years are from Thoburn (1977b, pp.95-111).
(63) In 1978 this varied from 5 per cent to 12.5 per cent depending on the size of profits.
(64) For the other two foreign operations it is difficult to give figures. Billiton's operations are included only as part of its parent company Shell, and Fairmont's include much property dealing which obscures the mining performance.
(65) Note though that petroleum products have been refined locally in Thailand since the 1960's, and it is the crude which is imported (Ingram, 1971, p.281).
(66) Gillis' estimates of RV must be necessarily with respect to changes in export revenue since tin production is not marginal to the economy. In other words RV for the tin export sector as a whole in Bolivia is not a valid measure of real income gains. Instead, as the previous subsection has shown, the gains to Bolivia from the existence of tin mining have to be estimated taking into account the effect of exchange rate changes on other sectors and on consumer welfare from imports.
(67) This figure makes only a very slight allowance for foreign ownership of capital in private sector mining in Bolivia.
(68) This uses the Squire-van der Tak (World Bank) methodology. See chapter six, especially the Annex.
(69) This is most likely to be true for onshore dredging in Thailand,

where project data were difficult to obtain.
(70) Although in this subsection and earlier in the chapter the export taxation of Bolivia has been discussed as if it consisted of a single export tax, it actually consists of an export royalty and (since 1972) an export tax. The former makes some allowance for presumptive costs and the latter for different grades of concentrate and (by very broad categories) different mining sectors. There is also some tax incentive for excess production. See Gillis (1978, chapter six) and Engel and Allen (1979, p.156).
(71) The effects of profits tax in South East Asia (it is not used in Bolivia) also may reduce supply, but its effect is very much less.
(72) A similar figure is not available for Thailand, though at the time of writing this study I understand that a 125 sector input-output table for Thailand is nearing completion.
(73) Thoburn (1977b, p.126) cites an earlier study by C. B. Edwards to the effect that petroleum refining in Malaysia has a very low ratio of domestic resources cost to foreign exchange saving. Thus the 'inappropriateness' of its high capital intensity would seem to be offset by economies of scale and advanced technology.
(74) Ingram (1971, p.284) describes Thai electricity production in 1950 as being 'pitifully inadequate', and the growth in supply has occurred since then.
(75) In 1965 (using an inverse input-output matrix for the 1965 table calculated by Lo, 1972, p.57) imports represented only 3.5 per cent of output value in the table of cumulative primary factors.
(76) The evidence in Malaysia is clear, however, that most engineering firms supplying the mining industries owed their establishment and early growth to mining even though they now may have a wider range of customers.
(77) This is relevant both to the reasons why the mining sector may prefer to buy locally rather than import, and to why the backward linkage industries did not set up initially as export industries in their own right.
(78) In 1973 approximately 0.9 per cent of manufacturing value added in Peninsular Malaysia was generated by the manufacture of tin cans and metal boxes and 0.8 per cent by fruit canning (Annual Bulletin of Statistics, 1976).
(79) This is not to argue that urbanisation is an unmixed blessing, as Lipton's (1977) work on urban bias has shown; simply that mining in remote areas is likely to generate less development than is mining geographically linked to the rest of the country.
(80) However, gravel pumping in 1972 in Malaysia employed per unit of output only a quarter of the workers on rubber estates and only 37 per cent of those on oilpalm estates (see Thoburn 1977b, chapter nine).
(81) See National Union of Mineworkers (1979). Their report also comments on the findings of an earlier NUM mission, in 1977.
(82) These poor conditions were also noted by the Harvard team several years earlier, though they felt things were worse in Comibol than in the private sector (Gillis, 1978, pp.38-39).
(83) Dredge figure from Monthly Bulletin of Statistics, Malaysia, July 1978 and the gravel pumping figure from Census of Mining Industries 1974. The Bulletin does not give a figure for gravel pumping.

# 6 The Distribution of Benefits from New Investment in Tin

## 6.1 INTRODUCTION*

Since the early 1970's much public discussion has taken place in the tin industry about the difficulties of new investment. From a peak in 1972 world production fell until 1976, and the fall has been particularly sharp in Malaysia. Malaysia output recovered somewhat in 1978 but was still (at 62,650 tonnes) below the 1972 peak. As chapter five has shown, the 1970's saw significant hardenings in attitudes towards foreign investment in Malaysia and Thailand, while some minor liberalisation has taken place in Indonesia. Both domestic and foreign investors in Malaysia have complained bitterly about taxation levels, not only in company reports but also in formal submissions to the government (1).

A case strongly argued by Malaysian dredging companies, including those of the Malaysia Mining Corporation, is that while existing operations remain extremely profitable even after over 25 per cent export duty and over 45 per cent tax on profits, new investment is discouraged by those tax levels. For gravel pump mines in Malaysia current profitability is adversely affected by taxes more than is the case with dredges, but the problem is rather more one of tax structure (in this case the reliance on tax on output rather than on profits) than of tax levels. The Thai mining taxation system, since it follows that of Malaysia closely, has similar effects on new mining ventures. In Indonesia, since few private investors are involved, and since the industry has been basically quite profitable at the price levels of the 1970's, taxation is less of a live issue.

Interest has been focussed on investment too as part of bargaining for the floor price in the International Tin Agreement, and the setting up of an Economic and Price Review Panel under the present (fifth) ITA. Bolivia particularly, as a high cost producer, has been dissatisfied with the floor price.

The investment appraisals presented in this chapter allow these issues to be considered in an informed fashion. First the private profitability of mining investment in the three South East Asian producers is discussed at prices ranging from the late 1979 ITA floor price of M$1500 up to M$2250, which was well above 1979 (and previous) price levels; and the effects of profit taxes and export duties are analysed. Some of the data for these appraisals were obtained by interviews in the region in 1978-79, though use was also made for Malaysia and Indonesia of very detailed feasibility studies produced by member governments for the ITC's Economic and Price Review Panel in December 1977 (2). Original data were not obtained for Bolivia to supplement the information on current profitability set out in the previous chapter. This is unfortunate but it has already been argued that the intra-marginal nature of most South East Asian production makes issues of taxation and foreign

TABLE 6.1
SOUTH EAST ASIAN TIN MINING PROJECTS

| | | MONTHLY THROUGHPUT ('000 CUBIC YARDS) (DS = DRY STRIPPING) | OPERATING DEPTH (FEET) | SIZE OF LEASE (ACRES) | YEARS OF PRE-PRODUCTION | YEARS OF MINING | CAPITAL COST (US$000) | GRADE OF GROUND WORKED - CONCENTRATE PER CUBIC YD | DATA SOURCE |
|---|---|---|---|---|---|---|---|---|---|
| MALAYSIA | | | | | | | | | |
| Dredges - | DA | 800 | 100 | 600 | 4 | 10 | 14,463 | 0.2 (0.121 kg metal/$m^3$) | ITC Model Study December 1977 |
| | DB | 800 | 100 | 850 | 4 | 15 | 14,565 | 0.2 | Extended Life version of DB |
| | DC | 1600 | 100 | 600 | 4 | 5 | 26,317 | 0.2 | Shortened Life version of DA |
| Gravel pumps - | GPA | 67+33 DS | 100 | 162 | 1 | 5 | 581 | 0.2 | ITC Model Study December 1977 |
| | GPB | 67+33 DS | 100 | 175 | 1 | 9 | 651 | 0.2 | Extended Life version of GPA |
| | GPC | 30+20 DS | 100 | 162 | 1 | 9 | 510 | 0.2 | Interview, Malaysia September 1978 |
| THAILAND | | | | | | | | | |
| Dredge - | DSH | 125 | 100 | 160 | 2 | 10 | 2,657 | 0.2 | Interview, Thailand November 1978 |
| Gravel pump | GPC | 15+3.75 DS | 30 | 80 | 1 | 9 | 425 | 0.4 (0.242 kg metal/$m^3$) | Interview, Thailand November 1978 |
| INDONESIA | | | | | | | | | |
| Offshore Dredge | DB | 650 | 100+ | - | 4 | 15 | 34,425 | 0.39 (0.23 kg metal/$m^3$) | ITC Model Study December 1977 |
| Gravel Pump | GPC | 20 | 30 | 85 | 1 | 9 | 226 | 0.4 | Interview, Indonesia April 1979 |

Notes
(1) Malaysia: DA - ITC Preproduction period shortened by one year; DC - two dredges in operation. Data on marginal capital costs of installing second dredge on existing site obtained from interview in Malaysia, Sept. 1978; GPA - ITC Preproduction period shortened by one year.
(2) Thailand: DSH - uses secondhand equipment; GPC - half-grade version also calculated.
(3) Indonesia: GPC - Could not cover operating costs at M$2250 at a grade of 0.2 Katis.

investment there especially interesting. Similarly Australia was excluded as were very minor producers such as Nigeria and Zaire. Data sources and methodology are described in more detail in an Annex to this chapter.

The issue of private profitability in relation to the present tax systems is important in countries with large private mining sectors in terms of securing future export revenue and in maximising tax proceeds without discouraging investment. Also important is the value of the investment returns from the point of view of society. The usefulness of social cost-benefit analysis, though not uncontentious, is widely accepted by developing country governments, and that therefore is the next step in the appraisals. The methodology used is that of Little and Mirrlees (1974) (LM) as extended by Squire and van der Tak (1975) (ST) for the World Bank. The main innovation of ST is to incorporate a system of weights reflecting local policies towards income distribution (both between groups of private individuals, and between them and the government), and they propose a three stage analysis. The first stage is financial (i.e. profitability to the private investor), the second is economic (which revalues inputs and output at 'efficiency' prices which reflect their true scarcity more accurately than do market prices); and the third is social, which incorporates weights reflecting local income distribution priorities.

For reasons discussed later, it is convenient to introduce the 'social' analysis at a later stage, while the 'economic' appraisals are used to investigate a number of other major issues:- namely the relative benefits of state, local, private and foreign ownership, and issues related to the choice of production technique.

## 6.2 PRIVATE PROFITABILITY AND THE EFFECTS OF TAXATION

### 6.2.1 Summary of projects

Table 6.1 gives a summary of the investment projects considered.
Malaysia: For Malaysian dredges, the ITC Model Study of December 1977 provides the basis. DA, a large capacity dredge is that of the Model Study, though its preproduction period is shortened by one year, allowing two years for exploration and site development and two years for construction. This shortening is more of an attempt to test the sensitivity of the ITC assumption than a suggestion of inaccuracy, (3) since lead times vary considerably between projects. It is also the case that much effort has been put into lead time reduction, resulting in a shortening of the construction period during the 1970's from 30 months to two years. This has been particularly so in Malaysia, where is is a result of potential competition from Japanese shipbuilders such as Mitsubishi, who would prefabricate parts rather than fabricate on site. Japanese competition however is much fiercer for offshore dredges (of which there are many in Indonesia and Thailand, and none in Malaysia), where a dredge can be constructed in a shipyard in eighteen months and towed to site. Mitsubishi in fact built the new Banka II dredge for Indonesia, Mitsui in Japan built the Billiton offshore dredge for Thailand, and the Billiton dredge for Indonesia was built in a Japanese-owned shipyard in Singapore.

DB is an extension of DA to a mining life of 15 years, on a slightly larger lease, which is more realistic and allows easier comparison with the Indonesia project. DC is an attempt to investigate an alternative pattern of resource depletion by putting in two dredges to operate the DA lease (with some resulting saving in initial capital costs). Virtually all new investment in dredging in Malaysia in the last ten years has used large capacity dredges of the DA type. The ITC study quotes the cost of the dredge in 1977 as M$25 million, and the total project costs (excluding privately purchased land) at nearly M$33 million. Figures obtained from local dredge builders and mining companies in Kuala Lumpur in 1978 suggest the dredge figure may be slightly high - in my interviews a range of M$19-25 million was quoted. The other project cost estimates seem reasonable. For example, for a dredge project completed very recently in Malaysia the cost of the dredge was 75 per cent of total project costs (excluding private land purchase), exactly the same proportion as in the Model Study.

For the Malaysian gravel pumping sector the ITC study represents a relatively large scale form of operation. Many much smaller mines are still being opened and the ITC versions are compared to data on a smaller private mine in Perak. GPA is the original ITC mine, and GPB gives extended life.

Thailand: as already mentioned access to the ITC study could not be obtained. This is particularly unfortunate since dredging investment data is almost impossible to obtain there. Only two new offshore dredges and no inshore dredges have been constructed since the Second World War. The first was built by Aokam more than ten years ago and its capital costs would be outdated, and the second is owned by Billiton Thailand, built in 1971. Since there would be no way of disguising its identity, Billiton were unwilling to release details either of the capital, or operating costs. In any case, it is a cutter suction dredge and therefore untypical, as well as being also rather too old for its capital costs to be a useful indiactor of present ones (4) (5). Thus the ITC study would depend on consultants' estimates for a large dredge of a type yet to be used in Thailand (though two large offshore dredges operate in Indonesia). Instead, a small capacity, secondhand dredge (DSH) was considered, based on information from a local mining company. This hardly provides a comparison with the dredging projects in Malaysia and Indonesia. Its use is mainly as an alternative to the gravel pump mine. Chapter four has shown that a number of local miners in Thailand have moved into dredging by buying secondhand equipment, which is rare in Malaysia. The small gravel pump mine (labelled GPC for comparison with Malaysia) is typical of Thai operations, which in both sectors are smaller and less technically advanced than in Malaysia. It would also have been extremely interesting to do an investment appraisal of the small suction boats which work off the coasts in the south west of Thailand, but their activities are illegal, or partially so, and data could not be obtained. However, the very useful study by Chulalongkorn University (1976) includes some rudimentary investment appraisals which are cited here.

Indonesia: here the dredge appraisal depends entirely on the ITC study, prepared by P. T. Timah. Apart from Koba Tin's small dismantleable dredges, no company in Indonesia other than Timah operates dredges, though Billiton in 1979 was about to start a large offshore dredging operation in Riau. Thus no independent data source was available (or really necessary). However, one is struck by the extremely high capital costs of DB in Indonesia (so-called because of its similarity in size to DB Malaysia). Compared to the Malaysian dredge cost of US $10.3 million (M$25 million), the Indonesian one costs US$23.3 million. Other project costs in Indonesia amount to approximately US$12 million, and in Malaysia (excluding private land purchase) to US$4.2. Nevertheless they do appear to represent the price Indonesia actually pays. The Belitung I dredge contract signed in early 1979 is for a dredge costing US$23.5 million, with total project costs of US$31 million, and the Bangka II dredge commissioned in October 1978 cost US$18 million, with total project costs of US$23 million (6).

The Indonesian gravel pump mine appraisal is based on figures supplied by P. T. Timah on Bangka island. It does not have the same degree of detail as the similar (GPC) gravel pump mines in Thailand and Malaysia. Only the total capital costs were given for an 'average' gravel pump mine (though some of the items included were listed and the total covered everything except prospecting). The capacity was unspecified, but since most such mine units on Bangka treat 15-20,000 cubic metres monthly, this is not a great problem, and 20,000 cubic _yards_ was assumed. The mine life was not specified and so ten years was assumed (including one year preproduction). Operating costs were assumed at roughly the mid-point of a narrow estimated range. Exploration costs on an assumed size site were taken, _faute de mieux_, to be the same as in Malaysia.

6.2.2 Comparison of private profitability across countries

The private profitability of new investment before income and output taxes is the basic indicator of the financial gains to be distributed between the various economic actors. It also serves as a reference point for international negotiations, particularly those concerning the ITA floor and ceiling prices. For these purposes, what is wanted is a comparison of profitability at the actual grades likely to be worked by new operations. Unfortunately data on the grades of actual investment projects cannot be obtained since this information is likely to be closely guarded by the participants, and _average_ grades on existing operations must often serve as a proxy. These still give a good general indication of the broad differences between countries and sectors however.

A further value of the data to producing governments is to compare technical efficiency between countries in similar operations. Tin mining is unusual even among mining operations in the extent to which capital and operating costs depend on _throughput_ (yardage) rather than the output of tin metal. Smelting, transport of the ore from mine to smelter, and of the metal to the consumer account for only a few percentage points of total costs, at least among alluvial tin producers. Thus in principle, say, the grades normally found in Malaysia (which are low) could be applied to Thai operations (where grades are higher) though some qualification will be shown later to be necessary to that view.

Table 6.2 shows the investment appraisal results in gravel pumping for the three South East Asian producers, and Table 6.3 gives similar figures for dredges. Reference back to Table 6.1 gives the grades used in the calculations, and it is worth discussing these before turning to the results. In Indonesia, officials in mid-1979 quoted grades for the two main mining islands as follows (one quintal equalling 100 kg metal per 1000 cubic metres).

| | GRAVEL PUMPS | DREDGES |
|---|---|---|
| BANGKA | | |
| Cut off grade | 2-3 quintals | 1 quintal |
| Average grade | 8 quintals | 5 quintals |
| BELITUNG | | |
| Average grades | 2 quintals | 1 quintal |

The cut off grades relate to prices prevailing at the time of M$1900-2000 a picul and the great difference between the two islands is clear, Belitung's average grades being at the cut off level on Bangka. In view of this wide dispersion, a grade of 0.242 kg metal per $m^3$ (0.4 kpcy concentrates) was chosen for gravel pumps, slightly arbitrarily, so as to be above the lower Bangka cut off and on the assumption that new mines would work poorer ground. For Thailand, the Department of Mineral Resources gives a grade of 0.4 kpcy (0.242 kg metal per $m^3$) as the average for the second half of 1977 for gravel pumps (7). This is used in the calculation and (conveniently) is the same as that assumed for Indonesia. For Malaysian gravel pumps no published data on grades are available. According to the submission by the All Malaya Chinese Mining Association to the Malaysian Ministry of Finance in August 1978 in connection with preparations for the 1979 budget, 80 per cent of the Association's members work on tailings or dredged- out ground, at a time when the latest published average grade for dredging was 0.19 kpcy. Accordingly 0.2 kpcy is taken as a rough approximation of the grade likely to be worked, and is virtually the same as used in the ITC Model Study (0.201 kpcy). For dredging in Malaysia the grade is also taken as 0.2 kpcy which, again, is almost identical to that of the Model Study (0.201 kpcy). For Thailand the DMR source estimates a range of 0.16-0.21 kpcy for 1977 for onshore dredges (8), so 0.2 again is taken. For Indonesia the ITC Model Study results are used, which assumes a grade of 0.23 kg metal per $m^3$ (equivalent to 0.39 kpcy) which is very close to the Indonesian and Thai gravel pumping grades.

Turning to the results in Tables 6.2 and 6.3 it can be seen that in Malaysia there is a noticeable difference between the profitability of the large scale gravel pumping operation based on the ITC Model Study (GPB) with the smaller and still common type of mine, GPC, though even at M$1500 GPC yields quite a respectable return. GPC in Thailand, a still smaller mine than GPC in Malaysia, but working on double the Malaysian grade, is substantially more profitable than GPC Malaysia (as might be expected), and at low prices it is much more so than its Indonesian counterpart (also on 0.4 kpcy). Some slight variation in results undoubtedly springs from the fact that GPB Malaysia's costs are for 1977, GPC Malaysia and Thailand for 1978, and GPC Indonesia for early 1979, but the influence of technical efficiency in GPB Malaysia versus GPC is clear, and the influence of higher grades shows in the Thai operations in comparison to those of Malaysia.

To compare further the technical efficiency with Malaysia a grade of 0.2 kpcy was also used in the Thai gravel pump appraisals. A similar procedure was not adopted for the Indonesian gravel pump operation

TABLE 6.2

CROSS-COUNTRY COMPARISONS OF GRAVEL PUMP MINING PROFITABILITY IN SOUTH-EAST ASIA - BEFORE TAXES AND EXPORT DUTY

| (US$000) | | | PRICE (M$ PER PICUL) 1500 | 1700 | 2000 | 2250 |
|---|---|---|---|---|---|---|
| MALAYSIA | | | | | | |
| GPB | NPV at | 5% | 2943 | 3941 | 5441 | 6690 |
| | | 10% | 2289 | 3099 | 4316 | 5329 |
| | | 15% | 1802 | 2474 | 3483 | 4324 |
| | IRR | | 78.71% | >100% | >100% | >100% |
| GPC | NPV at | 5% | 920 | 1921 | 3419 | 4664 |
| | | 10% | 567 | 1379 | 2595 | 3607 |
| | | 15% | 303 | 976 | 1985 | 2826 |
| | IRR | | 23.05% | 38.87% | 60.25% | 77.47% |
| THAILAND | | | | | | |
| GPC | NPV at | 5% | 4919 | 6947 | 9993 | 12532 |
| | | 10% | 3545 | 5193 | 7667 | 9731 |
| | | 15% | 2539 | 3908 | 5962 | 7676 |
| | IRR | | 41.99% | 55.08% | 74.80% | 91.02% |
| GPC | NPV at | 5% | <0 | <0 | <0 | 4649 |
| (Half grade | | 10% | <0 | <0 | <0 | 2046 |
| version) | | 15% | <0 | <0 | <0 | 144 |
| | IRR | | <0% | <0% | 4.39% | 15.45% |
| INDONESIA | | | | | | |
| GPC | NPV at | 5% | 616 | 2814 | 6113 | 8862 |
| | | 10% | 283 | 2065 | 4738 | 6966 |
| | | 15% | 42 | 1518 | 3732 | 5577 |
| | IRR | | 16.02% | 47.61% | 89.75% | >100% |

Notes

(1) Exchange rates are those of late 1977: US$1 = MS2.425 = 20.45Baht = Rp420.

(2) NPV's of all projects are adjusted to refer to a hypothetical 100,000 cubic yards monthly throughput (that of GPB Malaysia), for ease of comparison.

TABLE 6.3
CROSS-COUNTRY COMPARISONS OF DREDGE MINING PROFITABILITY IN SOUTH EAST ASIA - BEFORE TAXES AND EXPORT DUTY

| (US$000) | | | PRICE (M$ PER PICUL) 1500 | 1700 | 2000 | 2250 |
|---|---|---|---|---|---|---|
| MALAYSIA | | | | | | |
| DB | NPV at | 5% | 34801 | 45048 | 60415 | 73222 |
| | | 10% | 18172 | 24652 | 34370 | 42470 |
| | | 15% | 9078 | 13402 | 19888 | 25292 |
| | IRR | | 26.95% | 31.84% | 37.89% | 43.16% |
| THAILAND | | | | | | |
| DSH | NPV at | 5% | 6429 | 15113 | 28144 | 39002 |
| | | 10% | 1630 | 8226 | 18124 | 26372 |
| | | 15% | <0 | 3620 | 11353 | 17797 |
| | IRR | | 12.11% | 20.90% | 32.23% | 41.02% |
| INDONESIA | | | | | | |
| DB | NPV at | 5% | 56527 | - | - | - |
| | | 10% | 27443 | - | - | - |
| | | 15% | 11086 | - | - | - |
| | IRR | | 21.29% | - | - | - |

Notes

(1) Exchange rates as in Table 6.2.

(2) NPV's of all projects adjusted to refer to a hypothetical 800,000 cubic yards monthly throughput (that of DB in Malaysia) for ease of comparison.

(3) DB Indonesia results taken from ITC model study.

since (at 1977 exchange rates) the cost per cubic yard excluding depreciation was US$1.73 and is not even covered at a price of M$2250. However, since that particular set of data was collected in 1979, when the exchange rate had fallen from Rp420 to Rp625 to the US dollar, the procedure of using 1977 exchange rates for purposes of comparison for the gravel pump projects is open to some objection. At the lower exchange rate the costs per cubic yard including depreciation (9) are US$1.16, but still this is only just covered at a grade of 0.2 kpcy at M$2000. For GPC Thailand the change to the Malaysian grade leads to a negative NPV even at 5% at MS2000 and not until M$2250 is a reasonable return obtained.

Before concluding that the Malaysian operations are overwhelmingly technically more efficient, two caveats are necessary. First, Malaysian gravel pumps work at much greater depths than in Indonesia and Thailand. Most Malaysian gravel pump mines are at depths from 30 to 100 feet, whereas few mines in the other two countries are more than 30 feet deep. Since quoted grades are always an average of paydirt treated and overburden stripped, and since costs per cubic yard are lower for stripping, an overall cost per cubic yard (i.e. for ground treated and ground stripped) can be misleading. Second, and related, is that in a very rich mine of the sort still operating on Bangka, the operators would work the ground more slowly and carefully than on a poorer deposit, and again costs per cubic yard would rise.

In an attempt to isolate the causes of different levels of technical efficiency, some details on cost structures are given in Table 6.4, which includes details of other projects (GPA in Malaysia, DA and DC in Malaysia), essentially just variants on those already considered, discussed later (10). In Malaysia GPB has lower operating and capital costs per cubic yard than does GPC, and its labour use is less. Both operating and capital costs are higher on the Thai gravel pump mine whereas on GPC Indonesia higher operating costs are coupled with low initial capital investment, giving the project's profitability very great price sensitivity (11).

The relatively low operating and high capital costs of Malaysian dredges (except for the short-lived project DC) compared to Malaysian gravel pumps is evident, and the large dredge DB in Malaysia yields an IRR of nearly 27 per cent even at M$1500, the late 1979 ITA floor price. Unlike the gravel pumps, the rise in IRR's and NPV's as price increases is moderate. DB Malaysia yields a higher IRR than its Indonesian equivalent, despite the latter operating on ground twice as rich. Note though that the NPV's of DB Indonesia are scaled up to an 800,000 cubic yards per month equivalent, and if they are scaled back to their original level (for 650,000 cubic yards) only the NPV's at 5 and 10 per cent are higher for Indonesia. Note that the operating costs of DB Indonesia are more than a third as much again as DB Malaysia and the former's capital cost per cubic yard is triple.

The Thai dredge DSH is a secondhand, low capacity type, with high operating costs by dredging standards. Its NPV's and IRR's are of course very sensitive to the price paid for the dredge itself. An old dredge has very little value as scrap, so its price if it is sold for re-use on another lease is highly arbitrary, depending on the seller's assessment of what the buyer can pay. The price assumed is US$2.4 million (50 million Baht). Halving that price would almost double the

NPV at 10 per cent at the lowest price shown.

Finally, a note on the effects of the assumption used here that land in the Malaysia and Thailand projects has not been privately purchased. Estimates of the probable private purchase prices are given in the Annex. For Table 6.2 it is necessary to adjust the land costs to refer to the hypothetical 100,000 cubic yards monthly throughput on which all NPV's in the table are based. Thus US$433,000 should be deducted from the NPV's of GPB Malaysia, US$800,000 from GPC Malaysia, and US$800,000 from GPC Thailand, a large part of these differences for the same yardages and project lives being that the larger mine does not need a proportionately larger area in which to stack its tailings. Subtracting these costs from the NPV's leaves GPB Malaysia still viable at all prices and discount rates shown. GPC Malaysia, ceases to be workable at M$1500, but is viable at all discount rates by M$1700. The double grade version of GPC Thailand remains viable even at 15 per cent at M$1500, though the low profitability of the half grade version is further reduced, and even at M$2250 it is viable only at the discount rates of 5 per cent and 10 per cent. Note, however, that the land purchase price applies to the richer grades of ground normally found in Thailand, so for the half grade version land would probably be cheaper.

For the dredges US$2.1 million should be subtracted from DB Malaysia and US$1.9 from DSH Thailand. For DB this has little adverse effect on its basically high profitability, the project remaining viable at all prices and discount rates shown. DSH, however, is now made unviable at M$1500 except at 5 per cent, though at higher prices it can still function.

In practice, of course, whether private projects can function depends on tax payments too, and this question is now taken up.

### 6.2.3 The effects of taxation

The general structure of tin mining taxes has been discussed in the previous chapter. Taxes in Bolivia and Indonesia, where the bulk of mining is carried out by the state, are less of a live political issue than in the other producers. Nevertheless, they are important in securing to the government a larger share of the profits than it might get as the sole shareholder, given that the use to which profits are put by a state corporation may well be less socially desirable than the use to which the government would put them.

The major distinctive feature of mining taxation is that it is in principle possible for a government to capture the 'rent' element in the returns to mining investment. Tin is an industry where production costs and accessibility of deposits vary greatly between countries, and the world price normally covers the unit costs of Bolivia, the main high cost producer. A scarcity rent therefore accrues to the intra-marginal producers, principally those in South East Asia, over and above the return necessary to produce new investment. Thus it is not objectionable in principle that the mining industry in those countries should be subject to heavier taxation than, say, manufacturing.

As already noted at the start of the chapter, the level of taxation has been subject to the greatest amount of protest in Malaysia, both

TABLE 6.4
COST COMPARISONS OF SOUTH EAST ASIAN PROJECTS

| (US$) | | OPERATING COST PER CUBIC YARD | CAPITAL COST PER CUBIC YARD PER MONTH | CAPITAL COST PER CUBIC YARD OVER PROJECT LIFE | EMPLOYMENT PER 1000 CUBIC YARDS PER MONTH |
|---|---|---|---|---|---|
| GRAVEL PUMPS | | | | | |
| Malaysia | GPA | 0.42 | 5.81 | 0.09 | 0.57 |
| | GPB | 0.42 | 6.51 | 0.06 | 0.57 |
| | GPC | 0.61 | 10.20 | 0.09 | 0.80 |
| Thailand | GPC | 0.91 | 22.69 | 0.21 | 1.60 |
| Indonesia | GPC | 1.73 (1.16) | 11.3 (7.6) | 0.10 (0.07) | 1.30 |
| DREDGES | | | | | |
| Malaysia | DA | 0.30 | 18.08 | 0.15 | 0.22 |
| | DB | 0.30 | 18.21 | 0.10 | 0.22 |
| | DC | 0.29 | 16.45 | 0.27 | 0.22 |
| Thailand | DSH | 0.59 | 21.27 | 0.18 | 1.36 |
| Indonesia | DB | 0.42 | 54.50 | 0.30 | 0.45 |

Notes

(1) Capital costs are undiscounted. Discounting would reduce somewhat the capital intensity of the dredges in relation to the gravel pumps.

(2) The labour use figure for DB Indonesia refers to the similar Bangka II dredge commissioned in 1978, not to the dredge of the ITC Model Study from which the other figures were taken.

(3) Exchange rates same as in Table 6.2. The bracketed figures under GPC Indonesia are those at the 1979 exchange rate when the data was collected (Rp625 = US$1).

from the local Chinese sector and from the increasingly state controlled dredging sector. In addition to the income tax (40 per cent) and the development tax (5 per cent) levied on the profits of all business enterpirses in the country, tin mines pay a tin profits tax varying from 5 per cent to 12½ per cent and an export duty varying from 22 per cent of gross output value at M$1200 to 26-28 per cent over the range M$1500 to M$2500. In Thailand a 35 per cent business tax is levied on company profits (30 per cent if the company is listed on the stock exchange of Thailand). Corresponding to this lower level of direct taxation compared to Malaysia, export royalty is equivalent to nearly 27 per cent of gross output value at M$1500 rising to a maximum of 30 per cent just below M$2000. In addition, a business tax equivalent to 4.4 per cent of the gross value of output less smelting charges is levied. There are in addition in both countries import duties and taxes on inputs such as parts and petroleum.

In Thailand although some complaints are expressed privately at the tax levels, there is far more concern among foreign firms about the investment 'climate' following the revocation of the TEMCO concession in the mid-1979's and the activities of illegal miners on foreign firms' leases. Basically the difference between the Thai and Malaysian situations is that the Thai tin industry is not yet having to cope with the very low grades and great depths of the Malaysian deposits, so taxes on output rather than on profits are generally less of a burden. Export duty in Malaysia, as in Thailand, dates back to the previous century and has the enormous advantage of administrative simplicity, being collected by the smelter and impossible to avoid unless the miner smuggles his ore out of the country. For a miner, export duty is equivalent to a production cost. Its imposition is likely to lead to a lower level of output than would otherwise be the case, and to 'high-grading' - the failure to mine certain low grade parts of a deposit which may then be permanantly lost (12).

In the mid-1970's falling output in Malaysia became a matter of serious concern. 1977 was designated by the Ministry of Primary Industries as the year for the tin industry's sustained development, and many of the industry's problems were aired at the National Seminar on the Mining Industry in Kuala Lumpur in August 1977 (Thoburn, 1978b). In the budget for 1978, details of which were announced in the autumn of 1977, some relief was given to the industry. The Tin Profits Tax was changed from a 10 per cent flat rate to a sliding scale of 5 per cent, 10 per cent and 15 per cent with an exemption for small profits, the export surcharge (introduced during 1974) was consolidated into export duty at a slightly lower combined rate, and accelerated depreciation provisions were provided temporarily for capital expenditure for all industries (13). The government also took over from the tin industry the cost of buffer stock contributions.

In 1978 dissatisfaction with the tax system was still being felt and formal submissions were made to the Malaysian Ministry of Finance by the dredging and gravel pumping miners organisations - the States of Malaya Chamber of Mines and the All-Malaya Chinese Mining Association, respectively. While complaining generally about the heavy tax burden, these submissions made little complaint about the income tax on profits (or about the essentially similar 5 per cent development tax on profits), though improved depreciation provisions were called for (14). Both wanted export duty lowered however (or at least they wanted to have the export surcharge incorporated in the export duty to be abolished). They also were in favour of the abolition of Tin Profits Tax, which was levied on profits after a certain standard deduction, and does not take account of the size of the investment. In the event, the following '1979' (i.e. autumn 1978) budget did reduce the top rate of Tin Profits Tax from 15 per cent to 12½ per cent, but the basic reliance on export duty remains intact, and in 1977 export duty and surcharge made up about 75 per cent of the Malaysian government's tax revenue from tin mining (Thoburn, 1978b, p.33).

Table 6.5 sets out profitability figures after tax and export duty for the Malaysian projects whose before tax figures are given in Tables 6.2 and 6.3. Remembering that the costs of the large dredge, DB, relate to 1977, the table shows a rate of return of 12-15 per cent at the then range of prices, even though (at M$2000) the taxes have creamed off 87 per cent of the NPV at 10 per cent. The after tax IRR for DB at M$1500 (8.98 per cent) compared to 7.84 per cent in the ITC Model Study, which had a year's longer pre-production period, used straight line depreciation instead of the actual provisions of the 1967 Malaysian Income Tax Act and allowed for private land purchase (15). The large gravel pump mine, GPB, is left with much higher absolute profitability in terms of IRR after tax than is the dredge, even though the post tax IRR is only 20 per cent of that before tax. Interestingly the NPV at 10 per cent of DB is reduced by tax proportionately more than that of GPB. This is a reflection of the very high price sensitivity of gravel pumping compared to dredging, discussed at some length in section 6.4. It indicates that the tax structure, though bearing very heavily on less profitable operations, is not sufficiently progressive to offset the very sharp rise in GPB's relative profitability as price rises. On the small gravel pump mine, GPC, whose costs refer to 1978, taxation bears very heavily and its return is negative at M$1700 and under 10 per cent at M$2000. This heavy burden on a basically profitable operation is simply the result of export duty, and very little profits tax is paid even at M$2000. At M$2250 NPV at 10 per cent for GPC is only 10 per cent of the before tax figure.

Table 6.6 gives similar figures for Thailand. Here the export taxes are even heavier than in Malaysia, and they greatly reduce profitability. At no price within our range would a Malaysian grade be feasible for present Thai gravel pump operations. The high sensitivity of gravel pumping to price, however, allows good profitability at present prices (at 1978 costs) at Thai grades. The results for DSH, the secondhand dredge, are rather surprising, and show that an operation which was profitable before tax is now rendered unviable at present prices. Halving the estimated 50 million Baht paid for the dredge would however produce positive NPV's at our discount rates from M$2000. Since of course the scrap value of a secondhand dredge is negligible, secondhand prices could easily be adjusted. In fact this particular dredge was bought for 25 million Baht in the early 1970's, and the price was doubled in our calculation to take account of the increased cost of new dredges as an alternative.

In Indonesia a very different tax structure prevails, with the export duty taking only 10 per cent of gross output while corporation tax takes 45 per cent of profit (16). In Indonesia, the state tin corporation, P. T. Timah, which controls nearly 90 per cent of output, has the govern-

## TABLE 6.5
## EFFECTS OF TAXATION IN MALAYSIA

| (US$000) | | | PRICES (M$ PER PICUL) 1500 | 1700 | 2000 | 2250 |
|---|---|---|---|---|---|---|
| PROFITABILITY AFTER TAX AND EXPORT DUTY | | | | | | |
| DB | NPV at | 5% | 4048 | 7215 | 11950 | 15873 |
| | | 10% | <0 | 1324 | 4383 | 6899 |
| | | 15% | <0 | <0 | 321 | 2028 |
| | IRR | | 8.98% | 12.11% | 15.58% | 18.41% |
| GPB | NPV at | 5% | 346 | 696 | 1184 | 1564 |
| | | 10% | 188 | 469 | 862 | 1170 |
| | | 15% | 67 | 299 | 622 | 876 |
| | IRR | | 18.41% | 28.71% | 41.02% | 50.56% |
| GPC | NPV at | 5% | <0 | <0 | 181 | 647 |
| | | 10% | <0 | <0 | <0 | 360 |
| | | 15% | <0 | <0 | <0 | 144 |
| | IRR | | <0% | <0% | 9.42% | 19.24% |
| RATIOS OF PROFITABILITY AFTER AND BEFORE TAX AND EXPORT DUTY | | | | | | |
| DB | NPV at | 10% | <0 | 0.05 | 0.13 | 0.16 |
| | IRR | | 0.33 | 0.38 | 0.41 | 0.43 |
| GPB | NPV at | 10% | 0.08 | 0.15 | 0.20 | 0.22 |
| | IRR | | 0.04 | 0.12 | 0.18 | 0.20 |
| GPC | NPV at | 10% | <0 | <0 | <0 | 0.10 |
| | IRR | | <0 | <0 | 0.16 | 0.25 |

Notes

(1) Exchange rates as in Table 6.2.

(2) GPC NPV's are multiplied up to a hypothetical 100,000 cubic yards monthly throughput for comparison with GPB and Table 6.2.

TABLE 6.6
EFFECTS OF TAXATION IN THAILAND

| (US$000) | | | PRICES (M$ PER PICUL) 1500 | 1700 | 2000 | 2250 |
|---|---|---|---|---|---|---|
| PROFITABILITY AFTER TAX AND EXPORT ROYALTY | | | | | | |
| GPC | NPV at | 5% | <0 | 687 | 1795 | 2871 |
| | | 10% | <0 | 144 | 1044 | 1919 |
| | | 15% | <0 | <0 | 490 | 1215 |
| | IRR | | 4.39% | 11.62% | 20.90% | 29.10% |
| GPC (half grade version) | | | | | | |
| | NPV at | 5% | <0 | <0 | <0 | <0 |
| | | 10% | <0 | <0 | <0 | <0 |
| | | 15% | <0 | <0 | <0 | <0 |
| | IRR | | <0% | <0% | <0% | <0% |
| DSH | NPV at | 5% | <0 | <0 | <0 | 342 |
| | | 10% | <0 | <0 | <0 | <0 |
| | | 15% | <0 | <0 | <0 | <0 |
| | IRR | | <0% | <0% | 1.27% | 8.01% |

Notes

(1) For comparison with Tables 6.2 and 6.3, GPC NPV's are multiplied up to refer to a monthly yardage of 100,000, and DSH to refer to a yardage of 800,000.

TABLE 6.7
EFFECTS OF TAXATION IN INDONESIA

| (US$000) | | | PRICES (M$ PER PICUL) 1500 | 1700 | 2000 | 2250 |
|---|---|---|---|---|---|---|
| PROFITABILITY AFTER TAX AND EXPORT DUTY | | | | | | |
| DB | IRR | | 14.9% | 16.9% (M$1600) | - | - |
| GPC | NPV at | 5% | <0 | 359 | 1981 | 3335 |
| | | 10% | <0 | 76 | 1391 | 2487 |
| | | 15% | <0 | <0 | 958 | 1854 |
| | IRR | | <0% | 11.62% | 36.38% | 54.39% |

Notes

(1) NPV's for GPC are multiplied up to refer to 100,000 cubic yards a month, for comparison with earlier tables.

ment as its sole shareholder, and the shareholders' meeting decides on the dividend to be paid. Though tax and export duty date back to the early days of the corporation, the case for their continued existence is strengthened by the government's experience in the mid-1970's with the state oil corporation Pertamina. Pertamina invested lavishly not only in facilities for itself, but in hotels, an airline and a wide range of dubious projects. The export tax is the only sure way (leaving aside the possibility that ore may be exported illegally to avoid it) of the government's preventing a state corporation from over-investing in its own facilities. In a country like Indonesia where salaries tend to be low, non-monetary benefits such as good accommodation, sports facilities, travel facilities are an important element in employees' real income, and a certain level of such expenditure is certainly permitted. Obviously if such expenditure are counted as 'costs', liability for corporation tax is thereby reduced; but corporation tax does at least allow a large part of the corporation's surplus funds to be invested in projects approved by the government rather than at the corporation's discretion.

To show the effects of taxation on the Indonesian dredging investment, use was made of the ITC Model Study. Until the Billiton project came on stream late in 1979 taxation on dredging applied only to Timah's operations (none of which, incidentally, use contractors) (17). At the rather high grades found offshore in Indonesia, taxation leaves an IRR of 14.9 per cent, though this would be slightly lower if the ITC project's 43 per cent loan capital (paying interest at 12 per cent) was replaced by equity as in all our other versions. The IRR rises to 16.9 per cent at M$1600, the highest price used in the Model Study. The effect of taxation on gravel pumping, where many of the mines are operated by local miners under contract to Timah, is shown in Table 6.7. The high degree of price sensitivity of GPC Indonesia again is apparent, and a rise in price from M$1700 to M$2000 makes a large difference in profitability. This is particularly important since payments to contractors (discussed in the previous chapter) are much less price sensitive, so most of the gains from high prices accrue to Timah - which in turn gives the contractors an incentive to smuggle out part of their production.

So far this analysis of the effects of taxation has ignored the fact that land in Malaysia and Thailand may have to be privately purchased (see the Annex to this chapter). Substracting US$2.1 million from the NPV's of DB Malaysia to take this cost into account makes the project unviable at the discount rate of 10 per cent at M$1700, and at 15 per cent at M$2000 and M$2250. However, some allowance would be given in the depreciation provisions, though the cost-benefit programme has not been rerun to take that into account. Most likely DB would remain viable at M$2250 at 15 per cent, but certainly not at 15 per cent at M$2000, nor at 10 per cent at M$1700. Recent World Bank and UNDP sources (see the Annex) suggest rates of 7 per cent to nearly 14 per cent for the discount rate for private profitability appraisal. In constant price terms 14 per cent seems extremely high and is quite out of line with 1977-78 interest rates, (18) and even 10 per cent may be on the high side. Thus DB would certainly be viable at M$2000 and possibly (though only just) at M$1700 too. In any case, a mine might in practice buy only part of its lease privately and use state land for the rest, (as in the ITC model study, for instance where only 60 per cent of the lease for the dredge was privately purchased). For GPB private land purchase of the whole lease, neglecting the favourable effects of additional depreciation allowances on profitability, renders the project unprofitable at M$1500, and M$1700 at 15 per cent. Assuming a discount rate under 10 per cent, and taking into account the increased depreciation allowance (19), the project becomes viable probably at about M$1600.

GPC is affected very severely in Malaysia if it has to purchase its land, and is scarcely viable even at the lowest discount rate shown at M$2250. It had already been very heavily hit by taxation, and at the current (1979) tin price of about M$2000 it needed a very low discount rate before it is viable. For this project, it is worth noting that at 10 per cent at M$2000, the NPV of export duty paid was M$2.73 million (US$1.13 million). This is more than enough to have covered the cost of private land purchase (US$800,000) leaving a modest profit from which payment of taxes on profits could have been made.

Private profitability after tax is also substantially affected for GPC Thailand by the possible need to purchase land privately. Thus the project, which was previously viable at M$2000 even at 15 per cent, is no longer viable at 15 per cent (its land purchase cost being US $800,000). This is significant since the World Bank (Bruce, 1976,p.89) suggests a discount rate of 15 per cent for the appraisal of private profitability. As in Malaysia, of course, there is a considerable element of chance as to whether potential mining areas are located in lands held under private title - which then must be purchased - or on state held land which can be leased for a relatively low premium. DSH Thailand, the secondhand dredge, which has already been severely affected by taxation, is rendered unprofitable even at 5 per cent at M$2250. No doubt for this project a higher grade of ground would need to be worked, or else a lower price paid for the secondhand dredge.

## 6.3 ECONOMIC ANALYSIS OF PROFITABILITY

The aim of 'economic' appraisal, as opposed to the 'financial' (or private profitability) analysis of projects, is to use prices for the inputs and outputs which adjust for 'distortions' in the domestic economy. There may be, for example, different rates of import duty on machinery in different producing countries, and excise duties may vary on major inputs such as petroleum products. These differences were not allowed for in calculating the 'before tax and export duty' IRR's and NPV's in the previous section. Beyond those disparities are ones which arise where market prices do not reflect true scarcities, such that, for instance, interest rates are low in situations of relative capital scarcity, and wage rates for unskilled labour may be substantially in excess of the value of output foregone when such workers leave their previous activity. These distortions in developing countries are well documented (20). Since they often cannot be tackled directly, project selection procedures can be used such that approval is given to projects which move the economy towards a less distorted pattern of resource use (21). Further details of the procedure used are given in the Annex to this chapter.

Table 6.8 presents the results of the economic appraisals in the form of ratios between the <u>economic</u> NPV's and IRR's and the <u>financial</u> ones obtained in the appraisals of private profitability before tax and export duty for Thailand and Malaysia. Indonesian figures are not given since, for reasons explained in the Annex, it is thought that profitability before tax and export duty roughly approximates to economic profitability. For all the projects, economic profitability is consistently greater than financial profitability before tax. Basically this is because output is sold at the world price (22) and inputs are subject in many cases to excise taxes or import duties, while shadow wages are below market wages. Note that it is assumed for the individual projects that average export revenue equals marginal export revenue, that is to say that their increased output does not affect world price. This assumption is justified in the economic appraisals to the extent that the growth in output is designed to keep pace with the secular growth in demand for tin metal at given prices.

On all projects two further features stand out. First, the gap between financial and economic profitability lessens as the expected tin price rises. Since the reappraisal using efficiency prices in these cases only affects input prices, and since costs become a smaller influence on the cash flow as output price rises, this is to be expected. Similarly, as price rises, the differences between countries with respect to the economic; financial NPV and IRR ratios tend to lessen. The other common factor is that the gap between financial and the economic returns rises with the discount rate. An extreme case of this can be seen in the half grade version of GPC Thailand, where the IRR is only slightly in excess of 15 per cent, so the NPV at 15 per cent is near zero, with a résulting ratio of the economic to the financial NPV of over 45. Again, as in the influence of a price fall, a higher discount rate increases the influence of costs in the cash flow.

The influence of price can also be seen in the cross country comparisons, Thus GPC Thailand has consistently lower relative economic profitability than GPC Malaysia, because the latter (operating with a much lower grade) is considerably less profitable privately, and the different efficiency prices between the two countries do not cancel out this influence.

These results establish a general presumption that tin mining is more beneficial to the economy than the analysis of private profitability would suggest. So far, however, the results refer exclusively to state-sector projects. If the assumption is made that private capitalistic profits are likely to be invested and are no less valuable than uncommitted social income, the economic analysis can be used directly also to assess the economic desirability of different techniques of production. Further work can be done too to determine the benefits of allowing foreign investors to participate in tin mining. The next two sections discuss these issues.

## 6.4 THE CHOICE OF PRODUCTION TECHNIQUE

The major issue with regard to the choice of technique is that of the choice between the two main methods, gravel pumping and dredging. A similar question arises with regard to the use of offshore dredges or suction boats. This section also considers some alternative project time profiles in order to explore depletion issues.

TABLE 6.8

RELATION BETWEEN ECONOMIC AND PRIVATE PROFITABILITY BEFORE TAX AND DUTY - MALAYSIA AND THAILAND (RATIOS OF NPV's AND IRR's)

| | | | PRICE (M$ PER PICUL) | | | |
|---|---|---|---|---|---|---|
| | | | 1500 | 1700 | 2000 | 2250 |
| MALAYSIA | | | | | | |
| DB | NPV at | 5% | 1.22 | 1.17 | 1.13 | 1.11 |
| | | 10% | 1.29 | 1.22 | 1.16 | 1.13 |
| | | 15% | 1.44 | 1.30 | 1.20 | 1.16 |
| | IRR | | 1.25 | 1.21 | 1.20 | 1.17 |
| GPB | NPV at | 5% | 1.35 | 1.26 | 1.19 | 1.16 |
| | | 10% | 1.37 | 1.28 | 1.20 | 1.16 |
| | | 15% | 1.40 | 1.30 | 1.21 | 1.17 |
| | IRR | | >1.27 | Both >100% | Both >100% | Both >100% |
| GPC | NPV at | 5% | 2.40 | 1.67 | 1.38 | 1.28 |
| | | 10% | 2.91 | 1.79 | 1.42 | 1.30 |
| | | 15% | 4.09 | 1.96 | 1.47 | 1.33 |
| | IRR | | 2.25 | 1.80 | 1.59 | >1.29 |
| THAILAND | | | | | | |
| GPC | NPV at | 5% | 1.49 | 1.35 | 1.24 | 1.19 |
| | | 10% | 1.59 | 1.41 | 1.28 | 1.22 |
| | | 15% | 1.73 | 1.48 | 1.31 | 1.24 |
| | IRR | | 1.78 | 1.68 | >1.34 | >1.10 |
| GPC | NPV at | 5% | | <0 | <0 | 2.98 |
| (Half grade | | 10% | Both <0 | <0 | <0 | 4.90 |
| version) | | 15% | | Both <0 | <0 | 49.95 |
| | IRR | | <0 | <0 | 6.85 | 2.74 |
| DSH | NPV at | 5% | 2.26 | 1.49 | 1.23 | 1.44 |
| | | 10% | 5.07 | 1.76 | 1.31 | 1.19 |
| | | 15% | <0 | 2.45 | 1.43 | 1.26 |
| | IRR | | 1.82 | 1.44 | 1.27 | 1.22 |

Notes

(1) A negative figure indicates that the financial profitability was negative while the economic profitability was positive.

(2) '>' indicates that the economic IRR in question was >100 (beyond which the computer programme does not print out).

In an earlier paper (Thoburn, 1977a) I have shown that dredging and gravel pumping are competing techniques, insofar as many onshore dredging properties could in principle be operated by gravel pumps. Earlier chapters have shown that the two sectors in Malaysia and Thailand have been associated historically, with foreign companies and local (immigrant Chinese) miners, respectively; though this distinction in recent years has been broken down somewhat by an influx of local firms into dredging in Thailand and by federal and state government participation in the dredging sector in Malaysia.

Further work, which would be best done by specialist mining consultants, would be necessary to establish what would be the costs of a deep multiple gravel pump operation shifting the same yardage as a large dredge. Certainly there are gravel pump mines in Malaysia operating as deep as many dredges and there is one Malaysian gravel pump operation (Gopeng Consolidated, in Perak) which does work about 800,000 cubic yards monthly (23). While in a multiple gravel pump operation there would be economies of scale probably in ore treatment and the use of vehicles, supervision costs might rise disproportionately - and it is generally agreed that the workforce on a gravel pump mine needs very close supervision. Nevertheless some indication of relative profitability can be given by comparing the IRR's and the NPV's of the two methods, with the NPV's being adjusted so that each refers to the same monthly yardage.

Reference back to Tables 6.2 and 6.3 gives the before tax profitability figures. For Malaysia, adjusting the NPV's of GPB upwards to refer to a monthly yardage of 800,000, gives the result that GPB's NPV's are consistently higher at all the prices shown than those of DB at discount rates of 10 per cent and 15 per cent, and consistently lower at 5 per cent. This is true also of the after tax profitability position (Table 6.5). However DB has a longer operating life than GPB. If another dredge, DA, is used for comparison - which has a life of ten years - it is found that GPB has consistently higher NPV's even at a 5 per cent discount rate.

Comparing DA, whose returns are not shown in the tables, with the small Malaysian gravel pump mine GPC, shows that NPV's at all discount rates and prices used are higher for the dredging project, both before and after tax and duty. The IRR ranking switches, however, with that of GPC being lower at low prices and higher at high prices. The switch occurs at M$1700 before tax and duty and at M$2250 after tax and duty. A switch in relative IRR's with respect to a change in price is a possibility whenever the profitability is compared of two techniques with different time profiles of capital and operating costs. Specifically, the IRR ranking may well switch in favour of a technique with low capital and high operating costs <u>vis-à-vis</u> one with high capital and low operating costs as price increases. The former type is, of course, easily identifiable with gravel pumping (or small boat mining, of which more later) and the latter with dredging. The reason for the switch is that, at low prices, the high operating cost technique earns little or no surplus over operating costs, whereas the low operating cost technique does. At higher prices, although the latter's surplus is the larger it is divided by an even larger capital base, and the rate of return on capital employed increases more rapidly in the high operating cost technique (24). Although the profitability switch generalises easily from the return-on-capital-employed to the IRR criterion, the NPV's do not switch as a result. This is because the first project to emerge with a positive NPV establishes a lead on the other and this lead is never lost since subsequent rises in product price raise NPV by the same amount for each project (assuming that each produces the same quantity of output) (Thoburn, 1977a, p.45). In tin mining, however, NPV's may well switch, and again in favour of the high operating cost/low capital cost technique. This is because of the additional difference between the techniques, that dredging investment has a longer lead time (see Table 6.1) (25). Note however that there is also a tendency for higher prices to shorten optimal project life with given techniques. (See Appendix).

Comparing DA, GPB and GPC as alternative means of operating a given deposit, and assuming (heroically) that the increased costs of management and supervision in the multiple gravel pump operations are offset by cost reductions elsewhere in the mine, would lead to a choice of GPB since it consistently has the highest NPV, provided the discount rate were not much less than 10 per cent. As it happens, GPB has not only the highest NPV (at 10 per cent) but the highest IRR. In principle, a conflict between IRR and NPV could occur, as indeed it would if a choice between GPC and DA were considered. Here GPC at M$2250 (after tax and duty) has the higher IRR and a lower NPV (at all three discount rates). In this case the difference is a reflection of the fact that GPC requires a much smaller initial investment than DA, but that it earns a higher return on that smaller investment.

Since conflicts between NPV and IRR as investment criteria occur often in practice, and quite frequently in these appraisals, it is worth noting that most economists would recommend using NPV to choose between mutually exclusive projects, such as alternative means of mining a given ore body, once the relevant discount rate had been established. This is essentially because it is assumed that the supply of mining land is in short supply whereas investible funds can be borrowed on the capital market. Thus a firm (or government) would maximise its total discounted profits (i.e. NPV) by borrowing to finance a more capital intensive technique which had a higher NPV but lower IRR. Moreover, using a capital intensive technique, normally characterised by high capital and lower operating costs, insures more effectively against the risks of fluctuations in commodity prices. Thus at temporarily low prices, the labour-intensive technique may not be able to cover its operating costs and may encounter severe cash flow problems which the dredging investment would avoid. The cash flow problem is intensified by the high costs of reopening a mine which has been closed temporarily since, in South East Asia especially, a gravel pump mine would soon flood if not in use. Note however, that this is as much an argument for price and revenue stabilisation (26) - in practice, for strengthening the ITA - as for using capital-intensive techniques unless their NPV's are in fact higher.

Although NPV would normally be the relevant criterion, one particular circumstance would favour IRR (27). This is where the operation is seen as a high risk venture and the investor in question would prefer to commit a small capital sum. This is most likely for a foreign firm investing in a country which it regarded as politically unstable (e.g. where a communist takeover was possible or where nationalisation without compensation was likely). A higher rate of return on a smaller investment would be especially attractive for a firm which felt it had little choice but to invest in order to secure its own raw material supplies, though (as earlier chapters have shown) there is little vertical integration between tin users and tin mining companies. Therefore, although this chapter will use NPV as the main criterion, it should be remembered that foreign companies in particular cases might prefer IRR (28).

The Malaysian projects give a good opportunity to examine the economics of the choice between modern gravel pumping methods and dredging. GPB has proved itself superior in NPV terms at 10 per cent provided one can assume that operating costs do not rise disproportionately in a multiple gravel pump operation. Lowering the discount rate reduces GPB's advantage, mainly because DA, and other dredging projects, have a longer lead time, which means that their profits are more heavily discounted. A discount rate of 7 per cent would result in virtual equality of NPV's. This in fact is the rate advocated by a recent UNDP report on Malaysia, though rates of up to nearly 14 per cent have also been suggested. While 10 per cent seems a good compromise, this sensitivity to the discount rate must be borne in mind (29). Moreover, the advantage of a large dredge would increase greatly if the depth were increased below 100 feet. Triple stage pumping would soon be required for the gravel pump mine, whereas a dredge, by appropriate lowering of the paddock water, could mine to 180 feet. The advantage of dredging would also assert itself on a property large enough to support several dredges on a 10-20 year operation, since there are economies of scale in site development costs, provision of spares, etc (30). These comments are made from a financial viewpoint, however, and an appraisal of relative economic profitability is also necessary, which will be set out later.

With regard to the choice between gravel pumping techniques in Malaysia, GPB is clearly superior to the smaller scale method, GPC. Mines like GPC are common in Malaysia (and in Thailand and Indonesia) but their use is only for working small deposits. Indeed GPB is virtually a larger version of GPC but one which takes advantage of the scale economies in earth-moving, treatment, etc.

In Indonesia the two projects GPC and DB are not competing, since the dredge works offshore (where indeed most new dredging investment seems likely to take place in Indonesia). The choice of technique made by P. T. Timah onshore seems to be simply that dredges work large deposits and gravel pumps work small ones. Most of the onshore dredges are prewar and their capital cost is therefore only of relocation and renovation. With their lower operating costs compared to gravel pumps, such reconditioned dredges may well be more profitable than gravel pumping, but any new investment in onshore dredging should consider very carefully the alternative of multiple gravel pumping. It is worth noting too that experiments in Indonesia with small dismantleable dredges, both by P. T. Timah and by Koba Tin, have been unsuccessful, (31) whereas Koba's gravel pumps are very profitable, apparently.

Thailand provides an interesting opportunity to assess the profitability of dredging using secondhand equipment, often prewar and reconditioned, and which has been chosen by a number of miners who also own (or have owned) gravel pumps. DSH, the secondhand dredge, works a deposit three times the depth of GPC, and deals with the same grade as GPC's 'half grade' version, (0.2 kpcy concentrates). Before tax and export royalty DSH is a perfectly viable project while the half grade GPC only yields a positive IRR at a price of M$2000. Almost certainly the extra costs involved in deep digging would render GPC unviable (in its half grade version) at 100 feet, so DSH has absolute superiority. However, (as discussed earlier) it is hard hit by Thai taxation on its output, and it works the average onshore dredging grade of ground in Thailand. On richer ground of the sort worked by gravel pumps it would be a superior technique, though of course the supply of such dredges is strictly limited.

It is unfortunate that new data could not be obtained on which to base an appraisal of small boat mining - which has grown to such large proportions in Thailand since the mid-1970's - in comparison to offshore dredging, but much useful information is contained in a study in 1976 by a team from Chulalongkorn University's mining engineering faculty (32). First developed from converted fishing boats, the suction boats have been improved over the course of several years to dig to a depth of about 60 feet. Though many offshore dredges in Thailand can dig to double that depth, much of the available ground is workable by the small boats. They effectively took over the Southern Kinta lease in south-west Thailand, for instance. More of a constraint is that they can cope with overburden on the sea bed of only 1-2 metres, and they cannot operate at all in the stormy season of May to September. Many dredges cannot operate in storms either, but they sometimes have protected areas where they can be relocated. Prices for a suction boat vary from 200,000 Bahts (US$10,000) for a converted fishing boat to about 1.5 million Bahts (US$73,000) for a purpose built vessel mounted on steel pontoons. The Chulalongkorn study suggested that the older small boats had capacities of 3-10 cubic yards per hour (or 720 to 2400 cubic yards monthly) and might mine in a day 1-2 piculs of ore, with an employment of 10-12 men. Taking a rough median output gives a figure of approximately 6 workers per 1000 cubic yards per month, which compares to under one on the Billiton dredge, and is nearly four times the employment coefficient of GPC Thailand (Table 6.4). Taking 200,000 Bahts and 720 cubic yards monthly for the smallest boat gives an investment cost per cubic yard per month of US$13.6, slightly more than a small Malaysian gravel pump (see GPC Malaysia in Table 6.4).

The Chulalongkorn study contains investment appraisals based on the costs of a 'typical' small boat, assuming a two year project life after which the boat is sold; this short period being to allow for uncertainty (presumably as to whether this type of mining will be allowed to continue). The figures for rates of return do not allow direct comparison with the results of the other projects here, but roughly they are equivalent to an IRR less the rate of bank interest (33). At the tin price prevailing in April 1976 (M$1073 a picul) the cut-off grade was 1 kpcy, but the unweighted mean grade worked by the five small boats whose operations had been studied was 5.1 kpcy. As a result, the rates of return for four of the five ranged from about 200 per cent to 500 per cent, and the fifth had a return of over 50 per cent. Moreover, these are returns to boats selling their ore to the provincial authorities who deducted about 35 per cent of the value to cover export royalty (about 21 per cent) and handling and realisation charges. Also, the returns are those to the owner of a boat, who pays the workers half of the net proceeds of the sales of ore. For the ordinary workers on the boat studied in detail these payments were roughly three times the (1978) average Thai mining wage of 40 Bahts a day (and the foreman and the diver got more still).

This extraordinary profitability has been accompanied by an almost explosive growth in the number of boats, and complaints by the dredging companies and the offshore Mining Organisation that future dredging investment will be unprofitable because the boats are 'picking the eyes' from the deposits. What is interesting is whether the boats have a more permanent place in the industry once the richest deposits have been worked out. Their 1976 cut-off grade of 1 kpcy is roughly double that worked by the offshore dredges but it was arrived at by treating profit sharing wage payments as costs and the real tin price has risen by more than a third since 1976 (34). Larger boats in 1976 had cut-off grades of about 0.6 kpcy and technical developments have improved performance since then. In 1976, operating costs per cubic yard on the smaller boats were about US$1.3 on the basis of labour costs imputed at normal mining wages, roughly three times that of an offshore dredge (see DB Indonesia in Table 6.4), but at continuing high tin prices the boats appear to have an important role to play in the immediate future (35). They are also a way of spreading the benefits from tin finds considerably more widely among the local population than do dredging projects. Whilst they obviously reduce the profitability of future dredging, they return to the seabed the ore they cannot extract themselves (their recovery rates are low) in a form which is easily mined again in the future. With high rates of export duty there are continued incentives to smuggle ore, but the provincial authorities deduct only minimal amounts for income tax (0.2 per cent of gross output in the example cited earlier) and no doubt the illegal traffickers in ore also demand substantial discounts.

Moving from private profitability to economic appraisal, Table 6.9 so shows IRR's and NPV's at 10 per cent for all the Malaysian projects. Again the slight superiority of GPB over DB is clear, and again it disappears (though this is not shown in the table) if the discount rate is lowered to 5 per cent. GPB is unambiguously superior to GPC. The Thai results are not shown in this table, but DSH is unambiguously superior to the half grade version of GPC.

In Table 6.9 the alternative gravel pump project, GPA, is also shown. This has the same project life as used in the ITC Model Study, and GPB (which has been used in most of the comparisons here so far) is an extended life version of it. The effect of the extra life is shown in NPV's at 10 per cent for GPB which are about 50 per cent higher than those for GPA. This was also true of the before tax and duty versions (not shown), whereas after tax and duty at M$2000-2250 the NPV's at 10 per cent were two-thirds higher for GPB. At lower prices, the advantage of GPB was even more marked.

Like GPB, DB in Malaysia is an extended life version of DA, which has the original ITC Model Study's project life. Many mining companies in Malaysia in the 1960's and 1970's, such as Berjuntai, installed several dredges operating side by side. To assess the advantages (or otherwise) of a much more rapid rate of resource depletion, another project

TABLE 6.9

ECONOMIC PROFITABILITY OF ALTERNATIVE MINING TECHNIQUES IN MALAYSIA

| (M$ MILLION) | | PRICES (M$ PER PICUL) 1500 | 1700 | 2000 | 2250 |
|---|---|---|---|---|---|
| DA | NPV at 10% | 40.8 | 53.4 | 72.4 | 88.1 |
| | IRR | 33.57% | 39.26% | 46.78% | 52.64% |
| DA | NPV at 10% | 57.1 | 72.8 | 96.4 | 116.1 |
| | IRR | 33.57% | 38.62% | 45.47% | 50.68% |
| DC | NPV at 10% | 35.0 | 50.1 | 72.9 | 91.8 |
| | IRR | 29.64% | 36.62% | 45.90% | 53.22% |
| GPA | NPV at 10% | 39.2 | 49.7 | 65.4 | 78.5 |
| | IRR | >100% | >100% | >100% | >100% |
| GPB | NPV at 10% | 61.0 | 76.8 | 100.4 | 120.1 |
| | IRR | >100% | >100% | >100% | >100% |
| GPC | NPV at 10% | 32.0 | 47.8 | 71.4 | 91.1 |
| | IRR | 51.95% | 69.92% | 95.90% | >100% |

Notes

(1) All the NPV's, except DC, have been adjusted to refer to a hypothetical monthly yardage of 800,000.

(2) DC's yardage is 1,600,000 per month, since it is designed for comparison with DA; DC having half of DA's project life and double its throughput.

was run. DC is based on half DA's project life, but with two dredges to work out the property in five years of mining. In the private profitability appraisals (not shown) DC proved inferior to DA at all prices and discount rates used, the divergence being sharper after tax and duty. A switch in the profitability ranking (except in NPV at 5 per cent) was noted if price was raised to M$2500 in the before tax version. It was assumed that the very large difference in profitability between DA and DC after tax and duty meant that no such switch would occur there at a foreseeable price, but it was decided also to run the two projects in the economic appraisals. These show that at 10 per cent DC yields a very similar IRR at NPV at 10 per cent to DA, and is superior at M$2250 (though not at a 5 per cent discount rate). After only five years' operation, too, DC's two dredges would still be in good condition and would have potential resale value. If each dredge were sold at, say, M$5 million at the end of the project (i.e. nine years after the start) this would add M$6.45 million to the project NPV at 5 per cent, and M$4.24 million at 10 per cent. Neither would reverse the advantage of DA to the private investor after tax and duty even at M$2250, but in the economic appraisal DC would be given an edge at as low a price as M$1700. In practice the secondhand value of a dredge would depend on whether other deposits in the hands of private buyers were available to which the dredge could be transported without prohibitive cost. Alternatively, the dredges could be retained in public sector for use on subsequent projects as exploration resulted in the discovery of new deposits (36).

So far, in assessing the economic (i.e. as well as the purely financial) benefits of the various projects in this section and the previous one, it has been assumed that all projects are in the public sector. This is true, of course, for Indonesia, apart from three foreign firms' projects (and except that some gravel pump mines are operated there by contractors). In Malaysia, though the Selangor state government operates one wholly state owned project, most state and federal ventures have been jointly developed by government and foreign investors, while most of the gravel pump sector is run by local Chinese. In Thailand, the gravel pump sector is owned locally, as is most of the onshore dredging, though offshore it is foreign companies (some with Thai equity participation) who predominate. If local capitalists reinvest their earnings, a case could be made for regarding these as being as valuable as funds in government hands; so the economic analysis here could be applied to local firms too (37). For foreign investors further work is needed, and this is carried out in the next section.

## 6.5 FOREIGN INVESTMENT VERSUS STATE OWNERSHIP

Chapter five has shown that in Malaysia the state has decided it must participate in the equity of existing and of new foreign mining ventures. The share of government is normally 70 per cent (leaving aside the distribution of ownership between the federal government and that of individual states). In Thailand, foreign investors may not hold more than 49 per cent in new mining projects, and new foreign offshore tin mining ventures must have 51 per cent Thai equity ownership initially, rising to 60 per cent within five years (38). However, the Thai share can be held by private citizens and government ownership is likely to be confined mainly to the future activities of OMO, the Offshore Mining Organisation. In Indonesia, under the second generation contracts signed with the three existing foreign mining ventures, a certain proportion of shares must be sold to Indonesians within a specified period and in practice P. T. Timah takes a large block, though neither percentage nor the period is uniform between the projects (39).

In any case the relative benefits of a nationalised operation and a wholly foreign owned one are worth comparing in view of the existence of largely nationalised industries in Indonesia and Bolivia. Obviously the compairson is easier if both use similar mining methods. In Malaysia, the Malaysia Mining Corporation companies, controlled (though not wholly owned) by the federal government, do use essentially the same methods as would be employed by wholly foreign companies, whereas in Indonesia, P. T. Timah uses more labour in a given operation than would a private company.

A comparison is useful too of the intermediate option of state equity participation, given that it is actually used in Malaysia and Indonesia, because it is often argued that this is a poor alternative to a properly designed tax policy, provided a government stake is not necessary to induce the foreign investor to come in the first place. McKern (1976) shows, for example, that in theory increased government equity participation in a foreign investment project leads to a lower rate of return on investment to the government in relation to that of the foreign investor. This is because the government gets the tax proceeds anyway without contributing equity, so at zero equity its rate of return is infinite, and thereafter an increased equity stake and an increased share of the net revenues lowers the return, since the tax take is always in the numerator (40). Of course, this in itself is not a clinching case against participation if the government is searching for the best outlets for its funds and can earn a higher return by participation than in other outlets. Further, a proper consideration of investment criteria, as discussed in the previous section, leads one to consider net present values as well as the ratio of rates of return between the government and the foreign investor as used by McKern. The methodology used is adapted from that of Little and Mirrlees (1974, chapter eleven) for evaluating private sector investment.

The project chosen for illustration is the large dredging venture, DB, in Malaysia. The potential benefits are the discounted values of the positive net revenues in 'economic' terms (which include all tax and duty payments). For a wholly foreign project the capital costs are taken as zero, (41) and foreign profit outflows (taken as equal to profits after tax and duty in the 'financial' appraisal of DB) are subtracted from net revenue. For a partially government financed venture the government equity percentage of capital costs (valued at 'economic' accounting prices) is counted as the capital costs, and the government share of private after tax profits (which otherwise would have been remitted by the private firm in foreign exchange) is added to the 'benefits'. The wholly government owned venture is simply DB 'economic', already presented (see Table 6.9). Although it would have been possible (given the data collected) to work out the return to government in purely financial terms, it was felt that an 'economic' appraisal was of greater interest.

Table 6.10 sets out the results of these appraisals. Three features stand out. First, the IRR's at all prices rise monotonically with the degree of foreign participation. This is to be expected on a priori grounds (à McKern 1976), though it is possible that the more complicated investment time profile could have given perverse results compared to the simple return-on-capital-employed used by McKern. Second, and more interesting, is the sensitivity of NPV's to the discount rate. At the prices of M$1500 and M$1700 at 15 per cent NPV again rises monotonically with the degree of foreign participation. At M$1500 at 10 per cent and 5 per cent it falls monotonically. At M$1700 the beginnings of the third feature are apparent - the NPV's fall at 10 per cent and 5 per cent with the increase in foreign share up to 75 per cent, but rise again slightly at the 100 per cent foreign version. This is repeated at the higher two prices at 10 per cent and 5 per cent but at all prices the 10 per cent and 5 per cent NPV's with 100 per cent foreign ownership are less than with 100 per cent state ownership. At M$2000 and M$2250, however, even at 15 per cent this pattern appears but the NPV's are higher at 100 per cent foreign ownership than at 100 per cent state ownership.

This sensitivity both to discount rate and price makes for difficult decision making by government. Work by Bruce (1976) suggests an economic accounting rate of interest for Malaysia of 11.7 per cent, whereas Veitch (1977) proposes one of 8 per cent. Taking 10 per cent as a compromise leads one to suggest, at the 1977 costs on which DB is calculated, when the average tin price was about M$1700, that (assuming that the price of M$1700 will continue in real terms) the NPV for Malaysia would be highest with complete government ownership. However, if foreign expertise were required(and if it were difficult to obtain without foreign direct investment) only a 15 per cent foreign share (of the percentages calculated) would give a higher NPV than 100 per cent foreign ownership. If this is insufficient to attract foreign firms, then the next highest NPV is at a 30 per cent share, the level actually adopted in Malaysia. A lower discount rate would also lead to a 15 per cent (or, if necessary, 30 per cent) foreign share and both would be preferable to 100 per cent foreign investment. Only at a 15 per cent rate would foreign investment definitely be preferable on 'economic' grounds.

## 6.6 INCOME DISTRIBUTION AND EMPLOYMENT

In the previous three sections the project analysis has been 'economic'; that is to say it has been based on efficiency prices designed to adjust for domestic distortions. In addition to using its project selection procedure as a means of reducing allocative distortions, a government may wish to use it to produce a more equal distribution of income and to reduce unemployment. If the country is also short of investible funds, and the government cannot fully rectify the situation by fiscal policy, it may wish to use its project selection procedure to do so. In such a case uncommitted social income becomes more valuable than income in private hands. In a wholly public sector project, such income in private hands would predominantly take the form of wage payments, which would mainly be used for consumption expenditure. There would be also local profits on locally purchased intermediate products. Further details of the procedure used here are given in the Annex to this chapter. Briefly, to move from the ST 'economic' appraisal to

TABLE 6.10
ECONOMIC RETURN TO MALAYSIA OF
GOVERNMENT EQUITY PARTICIPATION IN PROJECT DB

| | PRICE (M$ PER PICUL) | | | |
|---|---|---|---|---|
| (M$ MILLION) | 1500 | 1700 | 2000 | 2250 |
| 100% FOREIGN OWNERSHIP | | | | |
| NPV at 5% | 89.1 | 114.0 | 151.4 | 182.5 |
| 10% | 55.2 | 70.9 | 94.5 | 114.2 |
| 15% | 36.1 | 46.6 | 62.3 | 75.4 |
| IRR | ∞% | ∞% | ∞% | ∞% |
| 75% FOREIGN OWNERSHIP | | | | |
| NPV at 5% | 92.6 | 111.7 | 140.4 | 164.4 |
| 10% | 55.6 | 67.6 | 85.7 | 100.8 |
| 15% | 35.0 | 42.9 | 54.8 | 64.8 |
| IRR | 69.92% | 76.95% | 87.89% | 95.61% |
| 50% FOREIGN OWNERSHIP | | | | |
| NPV at 5% | 96.1 | 117.1 | 148.7 | 175.0 |
| 10% | 56.1 | 69.3 | 89.2 | 105.9 |
| 15% | 33.9 | 42.6 | 55.9 | 66.9 |
| IRR | 48.05% | 54.10% | 62.79% | 62.24% |
| 30% FOREIGN OWNERSHIP | | | | |
| NPV at 5% | 98.9 | 121.4 | 155.3 | 183.5 |
| 10% | 56.5 | 70.7 | 92.1 | 110.0 |
| 15% | 33.0 | 42.4 | 56.7 | 68.6 |
| IRR | 40.33% | 45.90% | 53.71% | 59.42% |
| 15% FOREIGN OWNERSHIP | | | | |
| NPV at 5% | 100.9 | 124.6 | 160.2 | 189.9 |
| 10% | 56.8 | 71.8 | 94.3 | 113.0 |
| 15% | 32.3 | 42.3 | 57.3 | 69.8 |
| IRR | 36.45% | 41.70% | 49.02% | 54.39% |
| 100% GOVERNMENT OWNERSHIP | | | | |
| NPV at 5% | 103.0 | 127.9 | 165.2 | 196.3 |
| 10% | 57.1 | 72.8 | 96.4 | 116.1 |
| 15% | 31.7 | 42.2 | 57.9 | 71.1 |
| IRR | 33.57% | 38.62% | 45.47% | 50.68% |

the ST 'social' appraisal involves adjusting the shadow wage and using as a discount rate what Bruce (1976) calls the Social Accounting Rate of Interest (ST's 'Accounting Rate of Interest') in place of the Economic Accounting Rate of Interest used up to now. Adjusting the shadow wage allows account to be taken of whether (and to what extent) the additional consumption resulting from the project is a cost or a benefit. This depends on the social valuation of income in the hands of the government _vis-à-vis_ that of workers at the country's average consumption level, and on the valuation of the consumption of the particular group of income recipients _vis-à-vis_ people at the average consumption level. Analagously, the SARI takes account of the relative valuation of social _vis-à-vis_ private income, whereas the EARI in ST terminology is simply (in neoclassical terms) the marginal product of capital valued at border prices (ST, pp.27-29 and 75-76).

It has been convenient for expositional purposes to present the 'economic' and the 'social' appraisals separately. The issues of foreign investment and choice of technique have been discussed only in 'economic' terms. Basically the reason for this has been the prior finding that reappraisal in 'social' terms makes generally very little difference to the results of profitability in economic terms, whereas the absolute size of divergence between economic and private profitability is substantial (at least in Malaysia and Thailand).

The reason for the lack of impact of social reappraisal is mainly that the shadow wage bill even in gravel pumping is not large in relation to output value. In dredging at a price of M$2000 the economic shadow wage bill for unskilled labour in Malaysia is 0.6 per cent of output value (for DB) and 1.9 per cent in Thailand (for DS), while that for GPC Malaysia is 1.5 per cent and GPC Thailand 1.25 per cent. Social profitability is below economic profitability at a given discount rate since the conversion factors for the wage bills to alter them from 'efficiency' to 'social' prices are greater than unity (42). This is because mining wage levels are higher than those of workers on the average consumption level. Thus although workers may be drawn from poor groups whose increased consumption is accorded a high social priority, nevertheless their consumption increase is distinctly non-marginal and the ST formulation reduces its weight accordingly (ST, p.137). One objection to these results is that they take account only of direct employment in the revaluation of the shadow wage bill. This is unlikely significantly to have affected the results however, since the main inputs are purchased from industries which employ little labour. For example, according to the 1970 input-output table for Malaysia, wage and salary payments made up only 1.7 per cent of gross output value in petroleum, though there was a higher figure of 17.2 per cent in electricity. In non-electrical machinery the share was 17.6 per cent, though the particular industries supplying engineering products to the mines may be more labour-intensive than that (Thoburn, 1977b, p.198).

With regard to the discount rate, there is some difficulty of choice. Bruce (1976, p.89) suggests a rate for the financial appraisal for Thailand of 15 per cent, an EARI of 11.8 per cent and a SARI of 10 per cent, though other methodological assumptions give slightly lower rates for SARI and higher for EARI. He suggests rates of 13.8 per cent, 11.7 per cent and 8.0 per cent, respectively for Malaysia. Veitch (1977) suggests 7 per cent and 8 per cent for the Malaysian financial and economic appraisals, respectively, with no estimate for SARI. In both countries therefore EARI is more than SARI, which increases the NPV's of the social appraisals somewhat but the divergences of view about a rate suggest it may be safer to stick to 10 per cent for both the economic and social appraisals, bearing in mind the slight downward bias in the social results. Ratios of social to economic appraisals for the main Malaysian and Thai proejcts are given in Table 6.11 at a 10 per cent discount rate for the NPV's. For Malaysia, the similarity in results is striking, no ratio of social:economic NPV or IRR being less than 0.95. For Thailand for the high grade GPC project the ratios are similarly high. Only for the other two (much less profitable) projects does social profitability fall significantly, and this is the result of the cost component being already a large item in the gross revenues; and the difference diminishes sharply as price rises. Note that these generally small differences between the social and economic results could be swamped by using a slightly lower discount rate for the social analysis, as suggested by Bruce (1976), and thus raise social profitability slightly above economic profitability. In our projects the reappraisal of the shadow wage bill and of the discount rate tend to cancel each other out in their effects on NPV. A social analysis was not conducted for Indonesia because a suitably detailed cost breakdown was not available nor (at the time) a good set of social cost-benefit parameters.

Finally, a more direct view of the employment creating effects of tin production and investment can be taken. It should be stressed that these measures should not be 'added' to the social cost-benefit results already presented, since the latter take into account fully the (direct) social benefits of creating employment (and redistributing income) which the projects bring about. Nevertheless they are useful for explanatory purposes.

Reference back to Table 6.4 gives a first indication of employment in relation to production on each project. To use these figures, the employment statistics per cubic yard for Thailand and Indonesia (where operating grades are double those of Malaysia, except for DSH Thailand which is the same) should be halved (except for DSH). The position is easier to follow from Table 6.12 which shows ratios of employment to output and to investment. Table 6.13 allows a comparison with the overall (direct) employment intensity of the sectors in which the projects are found. In Malaysia, the lower employment intensity of the DB project in relation to Malaysian dredging overall (all of which is onshore) is a reflection of the fact that DB represents the newest, large-scale technology. It is not a reflection of the grade of ground worked since DB operates on roughly the average dredging grade for Malaysia. The sectoral average for gravel pumping in Malaysia is much closer to GPC than to GPB, reflecting the continuing importance of the smaller kind of gravel pumping operation. Again, GPB and GPC work on grades which are very similar to the average for their sector. In Thailand GPC appears to have an employment intensity similar to that of its small scale Malaysian counterpart. In fact this is a reflection of its higher grade of ground, and Table 6.4 shows its employment per unit of throughput is double that of GPC Malaysia. The sectoral employment figure for Thai gravel pumping is puzzling, especially since GPC Thailand operates on the grade quoted as the sectoral average by the Thai Depart-

TABLE 6.11
RELATION BETWEEN ECONOMIC AND SOCIAL PROFITABILITY
IN MALAYSIA AND THAILAND - RATIOS OF SOCIAL
TO ECONOMIC NPV'S AND IRR'S

| | | PRICES (M$ PER PICUL) | | | |
|---|---|---|---|---|---|
| | | 1500 | 1700 | 2000 | 2250 |
| MALAYSIA | | | | | |
| DB | NPV at 10% | 0.97 | 0.97 | 0.98 | 0.98 |
| | IRR | 0.97 | 0.98 | 0.99 | 0.98 |
| GPB | NPV at 10% | 0.98 | 0.98 | 0.99 | 0.99 |
| | IRR | -----------Both >100%------------- | | | |
| GPC | NPV at 10% | 0.95 | 0.96 | 0.98 | 0.98 |
| | IRR | 0.96 | 0.97 | 0.98 | Both >100% |
| THAILAND | | | | | |
| GPC | NPV at 10% | 0.96 | 0.97 | 0.98 | 0.98 |
| | IRR | 0.97 | 0.98 | Both >100% | Both >100% |
| GPC (Half grade version) | NPV at 10% | Both <0 | 0.32 | 0.85 | 0.91 |
| | IRR | <0 | 0.78 | 0.92 | 0.94 |
| DSH | NPV at 10% | 0.68 | 0.82 | 0.89 | 0.92 |
| | IRR | 0.83 | 0.90 | 0.92 | 0.94 |

TABLE 6.12
DIRECT EMPLOYMENT CREATION IN SOUTH EAST ASIAN TIN PROJECTS

| | | WORKERS PER 10 TONNES ANNUAL OUTPUT | WORKERS PER US$ MILLION OF INVESTMENT |
|---|---|---|---|
| MALAYSIA | DB | 2.02 | 12.08 |
| | GPB | 5.24 | 87.60 |
| | GPC | 7.35 | 78.42 |
| THAILAND | DSH | 12.5 | 63.98 |
| | GPC | 7.35 | 70.52 |
| INDONESIA | DB | 2.17 | 8.26 |
| | GPC | 5.74 | 110.53 |

Sources

See Table 6.1.

TABLE 6.13
SECTORAL EMPLOYMENT INTENSITIES IN TIN MINING, 1978

| | WORKERS PER 10 TONNES ANNUAL OUTPUT | | WORKERS PER 10 TONNES ANNUAL OUTPUT |
|---|---|---|---|
| MALAYSIA | | BOLIVIA | |
| Dredging | 4.62 | Underground | 6.58 |
| Gravel Pumping | 7.86 | AUSTRALIA | |
| THAILAND | | Underground | 1.27 |
| Offshore Dredging | 1.40 | Open cast | n/a |
| Onshore Dredging | 11.23 | UNITED KINGDOM | |
| Suction Boats | n/a | | |
| Gravel Pumping | 18.71 | Underground | 3.48 |
| INDONESIA | | | |
| Offshore Dredging | 2.39 | | |
| Onshore Dredging | 7.26 | | |
| Gravel Pumping | 8.35 | | |

Sources and Notes

(1) All figures from ITC, MSB, May 1979.

(2) Based on ITC production figures for first six months of 1978, doubled.

ment of Mineral Resources. Under-reporting of output is one possibility. Undoubtedly several thousand tons each year are smuggled from Thailand, but this is associated more with the semi-illegal offshore suction boats, and even if one adds 2000 tons to the gravel pump output figure, the coefficient falls only to 15.4. In spite of its high labour usage compared to Malaysia, GPC Thailand is still less labour-intensive than many other Thai gravel pump operations. The secondhand Thai dredge, DSH, which works the same grade as all the projects in Malaysia, is extremely labour-intensive vis-à-vis the Malaysia operations. Table 6.13 shows that it is typical of its sector, where grades are generally low and the dredges in use are almost all antiquated (43). The very low labour use in offshore dredging is partly a reflection of the newer equipment in use, but also a reflection of grades, which are 2-3 times greater on average than those inland (DMR, April 1978). Although DB Malaysia (an onshore dredge) and DB Indonesia, which works offshore, share the characteristics of being large-scale modern design dredges, employment per cubic yard in the Indonesian project is double that in Malaysia (44). No doubt this is partially a reflection of P. T. Timah's policy of maintaining high employment levels. For onshore mining in Indonesia grades are very much higher than in Malaysia and Thailand, especially on Bangka, the largest mining island. Grades have already been mentioned on Bangka of 0.85 for dredges and 1.35 for gravel pumps, (45) or (respectively) four and nearly seven times those for similar operations in Malaysia. Even taking account of the fact that rich deposits would be worked more slowly (i.e. with lower throughputs) the employment intensity for each type of operation is clearly greater than in the other producers. Employment intensities for Bolivia, Australia and the UK are shown for comparison.

The primary concern of this chapter, however, is with the effects of new investment, and Table 6.12 shows clearly how these differ greatly between the projects. For a million US dollars of investment 110 jobs are created in Indonesian gravel pumping compared to only 12 on a Malaysian dredge. The Thai secondhand dredge creates surprisingly many jobs per unit of investment.

## ANNEX TO CHAPTER SIX: NOTES ON THE INVESTMENT APPRAISALS

For the private (or 'financial') appraisals the method used was that of Little and Mirrlees in their 1968 OECD Manual. The LM private method is one of straightforward, commercial project appraisal and needs little elaboration. Data on costs were obtained from interviews in Malaysia in September 1978, Thailand in October 1978, and Indonesia in April 1979, and use was also made (see Table 6.1) of data from project appraisals prepared by Malaysia and Indonesia for the International Tin Council - the ITC 'Model Studies'. Tax and depreciation rates for Malaysia are basically those of the 1967 Income Tax Act, and export duty rates were obtained from the Malaysian Department of Mines. In Thailand the Department of Mineral Resources kindly supplied details of tax and export royalty rates, and depreciation allowances. Tax rates for Indonesian tin mining were obtained from P. T. Timah in Jakarta.

So as not to complicate further the already complicated analysis, with its wide range of projects and countries, several simplifications were made.

1 No projections were made about tin prices, but a range of realistic prices chosen instead.

2 Each project was evaluated on the assumption of a constant product price throughout its life.

3 No allowance was made for rising real wage or for differential rates of inflation affecting different inputs (equal rates of inflation are dealt with by assuming all calculations to be in constant-price terms).

4 An average grade of ground (see Table 6.1) was assumed for each project. The grade is not objectionable in itself, but two problems arise. First, the average for GP's includes the yardage of barren ground which is dry-stripped during mining operations (it excludes, however, dry-stripping during any relocation when the mine is not producing). This is necessarily included because dredges must also deal with such ground during their normal operations, when it is included with yardage treated. The other objection is that the average may conceal differences in grade between different parts of the mining lease. As Trevor Williams (1973) has shown, at positive discount rates the NPV of a mining operation will be increased if the richest grades are worked first (and thereby discounted less in the project appraisal). Allowing for this would be similar in effect to allowing for higher product prices in the early years of the project, but it is difficult to do in anything but an arbitrary and taxonomic way (and the analysis anyway shows sensitivity to the product price, so that point does not need proving).

5 It was assumed that all private projects were set up on state land. In Malaysia land may either be obtained from the state government or from private individuals or firms who will sell the lease. Prices of mining leases privately purchased in Malaysia varied (in autumn 1978) from M$5000 to M$10,000 an acre. Thus a large dredging company planning a project that year was paying M$5700 for its land, and a gravel pump miner suggested M$10,000 as an absolute maximum for small leases. The ITC model study assumed that 60 per cent of its dredging project was on private land purchased at M$6000 per acre, the rest being on state land. For DB in Malaysia the cost of land privately purchased at M$6000 an acre would be M$5.1 million (= US$2.1 million), which would be incurred in the first preproduction year and could be subtracted directly from the NPV's (46). For GPB and GPC in Malaysia, the land costs at M$6000 per acre would be M$1.05 million and M$0.972 million (US$433,000 and US$400,000) respectively. In Thailand land can be privately purchased and in autumn 1978 a miner estimated that mining land would vary in price from 15000 to 30000 Bahts per rai (US$1833 to US$3666 per acre), the former for a lease including barren sections. The land costs for GPC and DSH would then have been US$147,000 and US$293,000 respectively, using the lower land price since all mines need sections on which to dump tailings. In fact it was assumed that land was acquired from the Thai Forestry Department at 400 Baht per rai per five years. In general, the price paid for a private lease partly reflects the economic loss from output foregone if the land is suitable for (say) agriculture,

and partly the economic rent from mining, in which the owner of the land wishes to share. Obviously a decision rule to determine whether to go ahead with a mining investment should consider the alternative use(s) to which the land could be put. It seems better to do this, however, by comparing the NPV per acre at the chosen discount rate between the two (or more) activities (47) rather than arbitrarily subtracting a rough estimate of the probable land value in another use without doing a proper appraisal of that use. It is worth noting in this context that in Malaysia in 1978 young rubber land was selling at M$2500-3000 per acre, rising to M$3500 for young rubber ready for tapping (48). No estimates of private land purchase costs have been made for Indonesia since almost all mining is done by the state and the state corporation already owns much potential mining land.

The Squire and van der Tak ('ST') procedure used here for the economic appraisals is essentially the same as that of Little and Mirrlees ('LM', 1974), if the assumption is made in the LM system that additional income in the hands of the government is no more valuable than the increased consumption resulting from employment in the project in question. Both methods use as their numeraire that of uncommitted income in convertible currency in the hands of the government, such income being expressible in terms of local currency at the official exchange rate. 'Border prices' (i.e. f.o.b. export and c.i.f. import prices) are the major source of reference on which commodity accounting prices are directly or indirectly based, and commercial cash flows must be converted into the equivalent of freely convertible foreign exchange. This involves valuing tradeable goods at their f.o.b. export price gross of export duty (for exportables) and valuing importables at their c.i.f. import price net of tax and duty; while non-traded goods are split into foreign exchange and rewards to domestic factors.

Eor most projects in Malaysia and Thailand very detailed splits were obtained of operating and capital costs. Converting cost items into economic accounting prices (or 'efficiency' prices in ST terminology) was made very much easier by prior work carried out by World Bank and United Nations Development Programme economists for Malaysia and Thailand (see Bruce 1976 (49) and Veitch 1977). Of the major non-traded components of capital costs, dry-stripping and site clearing were treated as 'construction', and a 'construction conversion factor' was applied from Veitch. Transport and smelting, and minor non-traded items were converted using the LM standard conversion factor (50). The ITC Model Study for Malaysian dredging conveniently showed import duties as a separate item but for other machinery items the conversion factors for Malaysia and Thailand from Veitch (1977) were used.

The shadow wage rate (opportunity cost of labour x SCF) used for Malaysia was M$70 per month at 1974 prices, (51) which was then divided by the 1974 average unskilled wage in each mining sector from the (latest) Census of Mining Industries, Peninsular Malaysia, which also gave breakdown of the workforce by skill categories. The conversion factors thus calculated were 0.30 for dredging and 0.44 for gravel pumping. For Thailand a shadow wage of 8 Bahts a day was used, based on a 1973 (Bruce, 1976) estimate of 6 Bahts as the opportunity cost of labour, updated for inflation and adjusted by an SCF of 0.791. Applied to an average 1978 unskilled mining wage (obtained from interviews) of 40 Bahts, gave a conversion factor for unskilled labour of 0.176. The small scale secondhand dredge, operated (typically) by a company also owning gravel pumps, was assumed to pay the same wages in both types of operations, so the same conversion factor was used for each sector's unskilled wage bill.

For Indonesia, some cost benefit parameters were obtained from an interview with the World Bank mission in Jakarta in 1979, but they are quite rudimentary (52). Furthermore the data on operating costs for both the dredge and the gravel pump operation in Indonesia were not broken down into constituent parts. On the basis of the cost structure of existing dredges (many of which operate offshore and use diesel, and are very old and therefore untypical of new dredges) it appeared that the implicit subsidy on diesel in Indonesia (see below) and the effect of revaluing labour at a shadow wage (crudely estimated by the World Bank Mission in Jakarta as 50 per cent of market wage) very roughly cancel out, so that the before tax and duty profitability is close to the profitability at efficiency prices. Most gravel pumps use more diesel and more labour per unit of output so, lacking a detailed cost breakdown, a similar assumption was made there too. For capital costs the ITC Model Study on the dredge gave a detailed breakdown, but the dredge was imported more or less complete and most of its components were simply traded goods already (in the absence of import duty) valued at c.i.f. prices. Insufficient data on the capital cost structure of the gravel pump mine was available to revalue at efficiency prices.

The largest single non-labour input in tin mining is power, predominantly diesel for gravel pumps and offshore dredges, and electricity for onshore dredges. Although petroleum conversion factors were given in the publications cited, it is diesel which is the overwhelmingly most important component of thepetroleum products used (53).

In Malaysia the excise and import duty rates on diesel account for 20 per cent of its retail price to miners. Veitch (1977a) suggests a conversion factor for petroleum products of 0.67, but it was decided to use 0.80 instead. Interestingly Veitch's conversion factor for electricity is 0.79, so there is virtually no efficiency distortion in mines' choice of fuel. Similar calculations for Thailand gave a diesel conversion factor of 0.9, since the diesel duty in Thailand is only 0.22 Baht per litre (according to the Department of Mineral Resources). This compares to Bruce's (1976, p.89) conversion factor for electricity of 1.28, presumably indicating some element of subsidy in the market price. In Indonesia, a substantial net exporter of petroleum products, many petroleum products are sold at prices below the world level. It was estimated that diesel sold at roughly half of the world price (54). Since gravel pump mines and offshore dredges generally use diesel rather than electricity, a conversion for electricity was not attempted.

For the economic appraisals, the relevant discount rate is the Economic Accounting Rate of Interest - EARI (Bruce, 1976, p.38), which is in principle the marginal product of capital at border prices and in practice the rate of return on the marginal public sector project adjusted by the SCF. Unfortunately, and as discussed in the text, there are wide variations in the estimates for Malaysia and Thailand, and no estimate for Indonesia, so it was decided instead to show rates of 5 per cent, 10 per cent and 15 per cent, and to add in the text some discussion as to which would be appropriate.

Moving from 'economic' to 'social' appraisals involves changing from efficiency prices to shadow prices in the ST system. The shadow wage rate must be adjusted to take account of whether (and how much of) the extra consumption of workers on a project should be counted as a cost or a benefit. Basically this is a question of comparing the social value of their consumption to that of people on the average consumption level. In the ST system the ratio between these two consumption levels is d which in turn depends on n, the elasticity of marginal utility cost to consumption. Thus if n=0, d=1, and no greater value is placed on increases in consumption at low income levels than at high income levels. Additionally, income at the average consumption level must be compared to the value of uncommitted social income in the hands of the government, the ratio of the latter to the former being v. In an economy short of savings to finance investment v will have a value greater than 1. Thus a composite measure

$$'w' \ (= \frac{d}{v})$$

is determined, which compares the social value of additional project consumption to that of additional government uncommitted income. On the simplifying assumption that the workers' reduced leisure (as a result of employment on the project) has no social cost, the shadow wage rate can be written as

$$SWR = ma + (w - m)\ (B - \frac{d}{v})$$

where a is the SCF (conversion factor for output in ST terminology), m is the opportunity cost of labour, w the market wage (not to be confused with 'w'), and B the consumption conversion factor (ST, p.84). Assuming workers save no part of their income, w - m measures their increased consumption from the project. Only if

$$B = \frac{d}{v}$$

does the SWR = ma. In such a case the value of additional consumption on the project (at the accounting price B) is exactly equal to a marginal increase in uncommitted social income. Given n, d can be estimated both for marginal and non-marginal consumption increases (ST p.137). It was calculated as 0.37 for Malaysian dredging, 0.48 for Malaysian gravel pumping, and 0.645 for Thai tin mining. Given the values of v from Bruce (1976), the conversion factors needed to convert the shadow wage bills from efficiency prices to social prices were estimated at 2.43, 1.71, and 3.15, respectively; v being taken as 2.201 for Malaysia and 1.35 for Thailand from Bruce (1976). In all cases the changes in consumption were assumed to be non-marginal.

For the discount rate it is necessary to move from the use of KARI to the use of SARI (the social accounting rate of interest)(55), where EARI is adjusted to take account of the value of v over time (ST, pp.114-5). Bruce's (1976) estimates of discount rates for Malaysia conflict with those of Veitch (1976) and for Thailand Bruce gives several alternative estimates. Thus, again, it has been decided here to make calculations at discount rates of 5 per cent, 10 per cent and 15 per cent, which cover the Bruce-Veitch range. No suitable estimates are available for Indonesia at present.

NOTES

* Since this chapter was written (in summer 1979) various tax changes have been introduced in Malaysia. These are discussed in the next (and last) chapter, which also considers fundamental changes in Bolivian mining taxes announced in 1980. Also, a summary of the ITC investment model studies, from which some of this chapter's data are drawn, has been published (in June 1980) by the ITC - see Engel and Allen (1979 - sic). That summary includes some information on investment in Thailand, Bolivia and Australia, and this is introduced here wherever possible.

(1) For a statement of the complaints see the speech by Rahim Aki, the chief executive of the Malaysia Mining Corporation, to the State of Malaya Chamber of Mines in May 1979 (published in Tin, July 1979).

(2) Feasibility studies (or 'Model Studies' in ITC parlance) were made for the ITC of an onshore dredge and a gravel pump mine in Malaysia, an offshore dredge in Indonesia, a gravel pump mine and an offshore dredge in Thailand, and for underground mines in Bolivia and Australia. Responsibility for releasing these studies rests with the member governments, not the ITC, and I am grateful to the Chief Inspector of Mines in Malaysia and the Director of Planning in the Indonesian State Tin Corporation for permission to use their studies. It is very much to be regretted that similar permission could not be obtained in Thailand, though I was able to discuss the methodology of the study with officials in the Department of Mineral Resources in Bangkok. Since no field trip was made to Bolivia it was not possible to make the on-the-spot representations normally necessary to secure the study's release, and the study for Australia is based on the operation of a single private mine and therefore is confidential.

(3) It was not thought worthwhile to test the sensitivity to lead time by itself, since it is obvious that shorter lead time increases profitability; and several other changes were made in the Model Study assumptions. These were that the projects were taken to be financed 100 per cent by equity instead of the equity/loan assumptions of the model study that depreciation provisions were those in force in the countries concerned, not simply straight line, and that mining was not carried out on privately purchased land. The effects of introducing private land purchase are also estimated, however.

(4) There are also available two tin mining feasibility studies, one by a research student (Wit Satyarakwit, 1971) and the other by a team of consultants (Huntings Technical Services Ltd., 1974), the former based on estimates for 1971 and the latter for 1973. Huntings produced a simplified social cost benefit analysis (using a 10 per cent discount rate), whereas Wit's work is entirely in financial terms. Wit's dredging capital costs were an arbitrarily updated version of a set of figures supplied by the Tongkah Harbour Dredging Company for an offshore dredge, a member of the TASK group (see chapter four) whose most recent new dredge (Aokam 3) was built in 1967-8. Huntings' source is cited as a lecturer at Thammasat University in Bangkok, and much reference is made by them to Wit's work too. Huntings cited the capital cost for a sea dredge of 300,000 cubic yards/month capacity as 100 million Baht (US$4.9 million), an estimate which would now be quite out of date and unusable since 1974-5 saw a rapid increase in the cost of steel and other dredge components.

(5) The ITC study published after these appraisals were conducted gives total capital costs for a Thai offshore dredge as US$29.9 million. Project costs for a gravel pump mine are given as US$469,600 (Engel and

Allen, 1979, p.145).

(6) As noted above, however, the Thai offshore dredge in the ITC investment appraisal cost nearly US$6 million less than the Indonesian one.

(7) See DMR, April 1978. Note that 'Katis per cubic yard' (kpcy) refer to concentrates not metal content whereas kg per $m^3$ refers to metal ('Sn'). Since these measures are the accepted conventions in Malaysia and Thailand, and Indonesia, respectively, they are retained here.

(8) The offshore dredging grades are 0.42 - 0.5 kpcy.

(9) Estimated on basis of P. T. Timah figures at US7 cents per cubic yard over a 10 year project life.

(10) These projects are discussed in section 6.4.

(11) Price sensitivity is discussed further in section 6.4.

(12) These issues are discussed more fully by Bucovetsky and Gillis in Gillis (1978), chapter four.

(13) See FEER, 11 November 1977. The accelerated depreciation provision, which allowed an 80 per cent write off, was introduced for all industries for two years only. In view of its temporary nature it was not used in the appraisals here.

(14) The SMCM wanted capital allowances to be payable when the expenditure was made rather than when the equipment came into operation. The AMCMA wanted losses on one or more mines under a single ownership to be allowable against profits in non-adjacent mines under the same ownership.

(15) This is the version of the Model Study where capital is provided entirely by equity. However the ITC model study assumed that 60 per cent of the projects' land was privately purchased at a cost of M$6000 per acre. The gravel pump project, in contrast, was assumed in the ITC study to be on state land. Note that the Indonesia Model Study did not show an 'equity only' version of the results.

(16) For the three foreign companies which operate in the tin industry in Indonesia there are minor variations in the tax liability, discussed in the previous chapter.

(17) P. T. Koba Tin's operations on Bangka island do use one small dismantleable dredge, but it is predominantly a gravel pumping venture and the dredge is due to be phased out in the early 1980's.

(18) According to the Economic Report 1978-9 issued by the Ministry of Finance in Kuala Lumpur in October 1978, the commercial banks' prime lending rate in June 1977 was 7½ per cent and that obtainable on long term government bonds in the USA was 7-8 per cent.

(19) Depreciation allowances do not reduce liability to export duty, which at 10 per cent discount rate for GPC at M$2000 took (in NPV terms) almost twice as much as did the taxes on profits.

(20) See Sen (1975).

(21) This is not to say that the accounting prices chosen for the appraisal should be those which would prevail in a distortion-free economy, since the project must take account of existing distortions which are beyond its control to influence. See Squire and van der Tak (1975, pp.26-7).

(22) This abstracts from differences between the price at Penang and that on the LME, and from the fact that Indonesia sells some of its output on long-term contracts.

(23) This is on a tenacious clay deposit which a dredge would have difficulty in operating.

(24) For a formal demonstration of this proposition in terms of IRR see Thoburn and Camina (1980). Thoburn (1978a) gives a more intuitive explanation with a numerical example.

(25) For an illustration of this proposition see Thoburn (1977a, p.50, n.24).
(26) In a predominantly demand shift market, such as tin, stabilising prices would tend to stabilise export earnings too.
(27) This case is argued strongly by Merrett and Sykes (1973, pp.123, 213-7).
(28) They would prefer IRR too if they were short of funds (or borrowing capacity) in relation to their available investment outlets.
(29) See the discussion of appropriate discount rates given in section 6.2.
(30) The advantages of multiple dredge operations over a short period compared to a single dredge operation on the same lease over a longer period are discussed later.
(31) This is because it has proved impossible successfully to scale down the size of dredge bucket, and the small dredges with buckets below 5 cubic feet simply cannot deal with obstructions, dense decayed vegetation, etc. in the ground they are working. Timah's miniature dredges get only a third to a half of their rated capacity of 60-80,000 cubic metres per month. The corporation originally had seven dismantleable dredges, dating mostly from the 1960's. It now has four, of which one is a cutter suction dredge which also cannot deal with rough ground and achieves only a third of its design capacity.
(32) See Chulalongkorn University (1976) (in Thai). I am grateful to Professor Chadab of Chulalongkorn University, Bangkok, for providing a copy of this study, and for useful discussions on the current economies of small boat mining. Other information on this topic was obtained in interviews in Phuket in November 1978.
(33) The Chulalongkorn formula uses a 'Hoskold factor' which deducts bank interest (at 8 per cent) before arriving at the rough equivalent of an IRR, but some interest charges have already been included in operating costs prior to the calculation of annual net revenue (Chulalongkorn University, 1976, pp.25-26).
(34) The Penang tin price deflated by the 1971 based UN index of manufactured exports rose from 796 to 1976 to 1084 in 1978 (Engel and Allen, 1979, p.113).
(35) The great price sensitivity of high operating cost/low total investment cost operations has already been established in this section in the comparison of gravel pumping with onshore dredging.
(36) If many new deposits were available at a given time, public sector implementative capacity would most probably prevent their simultaneous exploitation. In any case, if a large producer such as Malaysia or Indonesia brought many new projects on stream simultaneously world price would probably fall and this would have to be taken into account in the appraisals.
(37) This statement is contentious for Malaysia, however, where the government is trying to redistribute income away from local Chinese to Malays. There is a risk in these circumstances that Chinese miners may export their profits.
(38) From text of speech on 'Mineral Development Policy' by the Thai Minister of Industry, 22-23 March 1978 in Bangkok, translated by DMR Bangkok, October 1978. The same restrictions do not apply to mining beyond 200 feet depth, where 100 per cent foreign ownership is permitted initially (though it must fall to 40 per cent within ten years), but with present technology in tin dredging such depths are not possible. Note that Billiton remains 100 per cent foreign owned since its mining operation is carried out under contract to the Offshore Mining Corporation.

(39) See chapter five for details.
(40) A brief algebraic statement and a further explanation of this proposition is given in McKern (1976, p.26-33). In practice the government's contribution would never be zero because there are administrative costs in negotiating with foreign investors and the government might also provide some infrastructure.
(41) This assumes that the foreign capital inflow has no adverse effect on domestic savings and investment. If a domestic firm could have undertaken the project, it is therefore assumed that it invests its funds elsewhere in the economy. The opposite case would be if an equal amount of domestic funding were remitted overseas, in which case the whole capital expenditure would be a cost to the economy.
(42) In Malaysia it is 2.43 for dredging and 1.71 for gravel pumping, and in Thailand for both sectors it is taken as 3.15. Note in this context that the Thai dredging project was operated by local miners paying similar wages to those they would pay in gravel pumping.
(43) Note that the dredge from which the DSH project data were taken actually operates at about half its rated capacity (which is over 20,000 cubic yards a month). This also is typical of the Thai dredging sector.
(44) The Indonesian employment figure is taken from the Banka II dredge, commissioned in 1978, since no employment figures were given in the source for DB Indonesia; but the two dredges are unlikely to differ greatly in their employment. (Indeed the ITC model study data from which DB costs are taken are probably based on Banka II anyway.)
(45) Converted from quintals Sn to kpcy (0.1 kg Sn per $m^3$ = 0.169 kpcy concentrates).
(46) Remember, however, that in the tables in the text the smaller dredge and gravel pump project NPV's are multiplied up to refer to the same throughput as their larger counterparts.
(47) This procedure is open to the objection moreover that in Malaysia and in Indonesia mining land is in short supply whereas agricultural land over the two countries as a whole is not.
(48) Interview with Bumiputra Chamber of Mines in Malaysia, September 1978.
(49) Bruce 1976 also contains a useful summary and reformulation of the LM-ST methodology.
(50) The SCF for Malaysia (0.873) was obtained from an interview in the Economic Planning Unit, Kuala Lumpur, September 1978.
(51) This was estimated on the assumption that mining workers would most likely be drawn from the urban unemployed, following the discussion in Veitch (1977).
(52) I understand from a contact in the World Bank in Washington in the summer of 1979 that a full set of cost benefit parameters for Indonesia should be available before the end of the year. Unfortunately that is too late for this project to use them.
(53) In Malaysia, according to the 1974 Census of Mining Industries, 97 per cent of 'petroleum products' used in tin mining consisted of diesel oil.
(54) This is estimated by comparing the Indonesian price (Rp 30 per litre) with the Malaysian price before tax and duty (M$1-10 per gallon).
(55) 'SARI' is Bruce's (1976) terminology. In ST it is simply referred to as ARI, the accounting rate of interest.

# 7 Conclusions

This study has traced changes in the structure of the tin industry with respect to the role of foreign investment and its alternatives of state enterprise and local private sector production. Development effects which have occurred through market forces have been considered, and policies in the four main exporting countries have been set out at length and their effectiveness evaluated, together with some comparative material on the smaller exporters. It has been necessary to distinguish between the effects of policies on the gains from existing tin production and the implications of those policies for investment behaviour and the longer term development of the industry.

## 7.1 RECENT POLICY CHANGES

The tin industry has provided an interesting range of government responses to foreign companies. The effects of policy on local producers has tended to be incidental in the sense that they are generally subject to the same tax regimes and mining legislation, though in Malaysia under the New Economic Policy announced in 1970, it is intended to increase the indigenous Malay share in income, wealth and employment not only in relation to foreigners but in relation to local Chinese too. Policy effects on local producers, though incidental, have not always been insignificant, however. This is especially true of tax policy.

Policies to increase the host-countries' development gains from tin exporting can be grouped under four headings:

1 Ownership restructuring of foreign firms.
2 Taxation.
3 Forward and backward integration.
4 The encouragement of wider development effects.

### 7.1.1 Ownership restructuring

In Bolivia outright nationalisation of the three largest tin companies (Patiño, Hochschild and Aramayo) occurred in 1952 following a revolution. The largest of these, Patiño, was in its origins an indigenous Bolivian enterprise, but it had shifted its headquarters abroad and had become multinational in its operations. The state company, Comibol, now produces about two-thirds of Bolivia's tin, the rest coming from private medium and small-scale mines. Also in the 1950's Indonesia ended the concessions of the Dutch company, Billiton, and later formed a state cor-

---

* A problem in writing this study has been the speed at which policies and economic structures change. As far as possible individual chapters have been revised to take account of the most recent events, but the present chapter has provided an opportunity for some additional updating (to September 1980). It is appropriate to express tribute here to the tin industry's excellent magazine, Tin International, which is invaluable as a source of information on the latest changes.

poration now known as P. T. Timah to produce the country's entire output. In Malaysia in the early 1970's, the national corporation PERNAS acquired control of the largest foreign mining group, the London Tin Corporation, in a stock exchange takeover. The new Malaysia Mining Corporation was then established in 71/29 per cent partnership with the international mining house, Charter Consolidated, which had tin subsidies in Malaysia. MMC now controls about a quarter of Malaysian tin output, though dispersed (and therefore non-controlling) foreign holdings in its operating companies still account (at most) for about 55 per cent of the equity. Thailand in 1975, following the earlier overthrow of military government, revoked the lease of the Temco company, also owned by Billiton, and followed this by the provision that all new foreign dredging companies should have 51 per cent Thai ownership, rising to 60 per cent in five years, and that the 60 per cent rule must be followed by existing foreign companies wishing to renew their leases. In 1980 the figure was raised to 70 per cent for companies established twenty-five years. In Nigeria and Zaire large equity stakes also have been acquired by the government in foreign owned tin mines.

These changes in Malaysia and Thailand must be seen in the light of previous developments. In Thailand there had already been a great decline in foreign involvement, such that only three foreign groups were involved by the 1970's, one of which was a subsidiary of the London Tin Corporation. This was the result of local (mostly ethnic Chinese) gravel pump miners buying up secondhand dredges from foreign companies who were leaving. This tendency of local miners to enter dredging directly was very rare in Malaysia, and unlike Thailand, much foreign investment had taken place in new large capacity dredges (all inland). In both countries dredging's output share had declined to about a third and foreign involvement elsewhere in the industry was minimal. The most recent government-induced increases in local ownership have been accompanied in Thailand by the formation of the government Offshore Mining Organisation and in Malaysia also by mining enterprises set up by the governments of several individual states in the Federation.

To some extent the desire for increased local ownership has been simply the result of nationalistic feeling, as in Bolivia and Indonesia. It appears also to reflect the belief that resource rents cannot be secured fully by taxation. The increases in retained value (or net foreign exchange contribution) secured by the tin exporting countries by ownership restructuring (and taxation) may have costs in terms of immediate losses in output and longer term falls in investment. The Bolivian and Indonesian nationalisations were followed by substantial falls in output for about ten years, due immediately to the outflow of foreign staff, and in the medium term to a lack of replacement investment. Thailand has seen very little foreign investment since the 1960's but its output has increased due to an explosive growth of offshore mining by small suction boats, producing now over a quarter of Thai output. Malaysian output fell markedly over the mid-1970's, and though this was blamed on 'penal' taxation by much of the industry there were also severe difficulties in securing new mining leases. This was primarily because individual states in the Federation have the virtually exclusive right to grant leases and yet had little incentive to do so since the bulk of tax and export duty goes to the federal government. The problem appears to be resolving itself by joint ventures between individual state mining organisations and foreign companies or the MMC. In Nigeria there have been large falls in output following the outflow of foreign technicians after ownership restructuring.

### 7.1.2 Taxation

Taxation policy has been especially important in the main intra-marginal and non-nationalised producers, Thailand and Malaysia, where profitability is extremely high and taxation is a major means of securing the resource rents generated by foreign owned production. Taxation policy is also relevant to state owned enterprises since the government is likely to feel it will make better use of the profits than would the enterprise. In Indonesia, where there are also large resource rents in tin, this issue has undoubtedly been influenced by the lessons learned by the government from the behaviour of the state oil company, Pertamina, which squandered considerable sums of money in unsuccessful investments. In Malaysia and Thailand local companies have been subject to essentially the same taxes as foreign companies, in order to raise revenue, and the taxes levied on the three foreign companies now in the Indonesian tin industry have been broadly similar tothose paid by P. T. Timah. In Bolivia, the main marginal producer, taxation has been levied on Comibol such as to cause accounting losses in some recent years. In general, the project has not tried to examine 'fiscal linkages' - the use to which the governments put their tin mining tax revenue - at any length. It could be objected that this neglects the role of the state or naïvely assumes that government revenue automatically benefits the economy. In Thailand, Malaysia, Indonesia and all the minor exporters, tin now in no sense dominates the economy, and it would be difficult (and largely meaningless) to try to trace the use of tin revenue. In Bolivia tin influences general development much more, and where possible a wider view has been taken.

Historically, the governments in Thailand and Malaysia have depended on export duties or royalties for their revenue from tin. Simple taxes on exports have been administratively easy to collect and difficult to evade unless the ore was smuggled out of the country. In the late nineteenth century in both countries the industry was encouraged in order to provide revenue. In colonial Malaya export duties averaged 12-15 per cent of export revenue from the late nineteenth century until after the Second World War. In Thailand export royalty stood at 10-16 per cent from the late nineteenth century. Malaysia since 1948 has supplemented the export duty with a tax on company profits (at rates which have risen from 20 per cent to 30 per cent in 1951 nad 40 per cent in 1959). Since 1967 there has also been a 5 per cent development tax and (since 1965) a special tax on tin profits now ranging from 5-15 per cent. Company tax is levied in Thailand too, at a rate of 30-35 per cent and company taxation dates from after the Second World War. In Bolivia a complex system of income and other taxes was replaced by taxes on output in 1965. In the mid-1970's Malaysian export duty rates were raised with a surcharge to rise to 27 per cent of output value, and Thai export royalty rates could be levied up to a maximum of 30 per cent.

The structure of taxation is virtually as important as its level. Taxes levied on the value of exports (which are effectively taxes on output since very little tin is consumed locally in the main exporters) cannot make allowances for differences in costs between individual companies. In Bolivia the export royalty has been based on the difference between the gross value of output and presumptive costs and the latter only have allowed for broad sectoral cost differences. Similarly the export tax in Bolivia has varied according to the type of mine (and grade of concentrate). In Thailand and Malaysia it has been levied at the same rate on all sectors. In Bolivia in the late 1970's export taxation was accounting for over 20 per cent of output value and in Malaysia and Thailand nearly 30 per cent. In times of low prices, such taxes bear very heavily on mines with relatively high operating costs, like the local gravel pumping sector in Malaysia and Thailand, and yet they have not been sufficiently progressive to reap really large shares of profit for the government when prices have been high. Moreover, a tax system based almost entirely on output taxes, as in Bolivia, cannot effectively give tax rebate for investment expenditure or expenditure on exploration or mine safety (1). In general then taxes on profits are much to be preferred to taxes on output, provided compliance can be secured and provided both the government and the mines have the more sophisticated administrative and accounting system required for income taxation.

In fact, Bolivia in 1980 has announced a complete shift away from export taxes to a single flat rate profits tax of 53 per cent (2). Malaysia in November 1979 announced a revision of its export duty rates, such that the minimum tin price at which export duty was paid would be based on an estimate of actual mining costs and would be more progressive, rising to a maximum of 50 per cent. The mining costs would be reassessed at regular intervals and appear to be based on those of the (higher operating cost) gravel pump sector. The other Malaysian taxes have been retained, and the tin profits tax was at the same time raised from a maximum rate of 12½ per cent to 15 per cent. These changes in Malaysia make the taxation structure slightly less onerous for local miners, but are a long way from a system which takes true account of profitability. In Thailand export royalty remains in its crude form, though still supplemented by (rather lower) rates of profits tax than in Malaysia. In Thailand the opportunities for moving to a profits based system are greatly inhibited by problems of tax compliance in much of the economy and the books of small mining companies are difficult to check. This is true to some extent in Malaysia too, and in the smaller mines in Bolivia.

In Indonesia an export royalty of 10 per cent is levied on P. T. Timah together with a profits tax of 45 per cent and the distribution of the remaining profits between the government and the company is controlled. The retention of some tax on output seems appropriate in that state companies have some incentive to use profits to provide their officers with generous salaries or (if these are limited) with generous amenities. Taxes on the three foreign companies are very similar and appear preferable to the Thai and Malaysian arrangement. Provision is also made for the Indonesian government gradually to take up equity in profitable foreign ventures. Tax compliance is a problem in Indonesia to the extent that there has been considerable smuggling of tin ore from the country. Payments to Timah's contractors, which take inadequate account of price changes, may encourage this. In Thailand smuggling has been a problem in the suction boat sector but the semi-legalising of this kind of mining has brought an increase in the ore sold to the provincial authorities in the south.

In Bolivia a Harvard team has pointed out how tax policy is limited by the sheer importance of tin as a source of revenue to the government, when other sectors are very lightly taxed. Taxes on Comibol and the private mines have been so high as to discourage even replacement investment yet mining tax reform depends for its success on effective taxation being levied elsewhere. However, the shift to profits taxation is undoubtedly an improvement provided it can be implemented administratively. No such problems exist in Malaysia or Thailand where tin mining (or any other mining) is not of great importance in total revenue. In Indonesia, as in Nigeria, Australia and Zaire, other minerals are important even though tin is not, and tax provisions which differ greatly between minerals can give rise to dissatisfaction among foreign investors.

### 7.1.3 Backward and forward integration

In alluvial tin mining the main intermediate purchases are of power and engineering products. In Malaysia a substantial light engineering industry has grown up to service the demand of the tin industry for spares and capital equipment. About 20 per cent of the value of a Malaysian tin dredge and about 40 per cent of the capital cost of a gravel pump consists of local purchases. This industry has developed through market forces and policy has had little influence. In Thailand there are also engineering workshops to service the industry and their output has greatly expanded with the growing use of suction boats which are locally assembled. No quantitative information is available, however, on the Thai industry. It has received some help in the form of a light import duty on engineering products. In Indonesia extensive facilities exist for the rehabilitation and repair of mining equipment, dating from the Dutch period, but new offshore dredges are bought from abroad nowadays, from Japan. Japanese innovation in particular has been instrumental in shortening the construction period but Malaysian dredge contractors seem to be holding their own by fabricating on site. In Bolivia most mining equipment is imported.

Although qualitatively similar in their generation of backward linkages, the local gravel pump sectors in Malaysia and Thailand are more materials and power-intensive (per unit of output) than is dredging. With regard to power, onshore dredges tend to use electricity and offshore dredges and gravel pumps to use diesel. Malaysia and Indonesia are net petroleum exporters and Thailand has its own refining facilities. In Malaysia the tin industry has been a large enough consumer of petroleum products possibly to have influenced the establishment of local petroleum refining in the 1960's with government incentives. Before the Second World War tin was such a large consumer of electricity that its demand may well have brought about electricity production at an earlier state than would otherwise have occurred. These points are quite conjectural, of course, and do not apply to Thailand and Indonesia in part because of the geographical remoteness of the tin mines in these countries from the main centres of economic activity and partly because tin there was much less important (in GDP and export revenue terms) than it was in colonial Malaya.

Forward integration into smelting has occurred in all three South East Asian exporters, and Bolivia at last is acquiring sufficient domestic capacity to smelt its entire output, as have Zaire, Brazil and Nigeria. Australia still ships some concentrates to Malaysia for smelting. Malaysian smelting dates back to colonial times, and its establishment was aided by a prohibitive duty on the export of tin ore. The Thai smelter is owned 100 per cent by Billiton, and was set up in the 1960's, and P. T. Timah established smelting self-sufficiency in the 1970's. Almost all Thai output is also marketed by Billiton, and Timah markets the Indonesian output. International trading companies are still active in marketing Bolivian and Malaysian tin, but the Malaysia Mining Corporation now organises its own marketing. Of the LDC tin exporters only Brazil has a flourishing tinplate industry. It is significant that Brazil is a major steel producer with a large domestic market since tinplate manufacture is more appropriately an adjunct of domestic steel capacity than domestic supplies of tin metal. The Thai tinplate industry works far under capacity and its viability is threatened by imports. The entire Thai domestic market for tinplate would only absorb 3 per cent of the country's tin metal output anyway. Malaysia and Indonesia are planning domestic tinplate manufacture. Policy in Thailand and Malaysia towards tinplate has been limited to investment promotional privileges. In Indonesia prospective tin plating is linked to a steel project. Solder manufacture to meet the demands of the electronics industry in Malaysia and Thailand is also a possibility to use more local tin, but a recent consultants' report on forward integration (Lim, 1980) was pessimistic in general about further tin processing in LDC's especially for export.

As well as aiding in the establishment of smelting and tin using facilities, producer governments have also been concerned with the ownership of those facilities. The Bolivian and Indonesian smelters are owned by the state, though the latter is administered by the government tin mining organisation while the former is operated by a separate body. In Malaysia one smelter is locally owned and the other is foreign owned. The Malaysia Mining Corporation has been putting pressure on the smelters to allow it an equity stake by threatening to build a third smelter. The Thai tinplate industry is a joint Thai-Japanese venture, and the Malaysian one is also a joint venture with a Japanese partner, with the Malaysia Mining Corporation also having a stake. Of the backward linkages, only petroleum refining has significant foreign ownership and only in Malaysia and Thailand.

### 7.1.4 Wider development effects

Information on local market creation is available for Malaysia where about 70 per cent of workers' expenditure is made locally. The Bolivian figure is very similar.

Gravel pumping has an employment intensity per unit of throughput roughly double that of dredging and the difference is more pronounced per unit of investment (though GP capital equipment is shorter lived). Suction boats generate at least six times the employment per unit of throughput compared to offshore dredging (though the figure varies greatly from boat to boat and there are no data to calculate a sectoral average). Indonesia appears to have made a conscious decision to operate its tin mining more labour-intensively than its Thai and Malaysian neighbours.

Only in Bolivia is the tin industry large enough to affect significantly the distribution of income in the economy as a whole, and there is no differential in favour of mining wages vis-à-vis manufacturing. On the contrary, many reports note the very low living standards of tin mine workers. In Asia mining wage differentials in the local sectors are slight in relation to manufacturing, and even in dredging Malaysian wages are only 50 per cent higher than in manufacturing. Real living standards appear to be significantly higher than in Bolivia reflecting the probability that workers are receiving a share of the mineral rents in Asia. In suction boat mining in Thailand the share of workers in profits is greater still.

In Asian tin mining 30-40 per cent of the workforce is skilled, and the skills are generally usable elsewhere in the economy. In the long run this is a force making for an improvement in the quality of the labour force.

## 7.2 THE DEVELOPMENT GAINS FROM EXISTING PRODUCTION

Increases in retained value through the restructuring of equity ownership and through taxation and backward integration are methods of increasing the net foreign exchange contribution of tin exports (3). Increases in the net foreign exchange contribution can be brought about also by the host-country securing a larger share of tin products' final consumer value by forward integration and by ownership restructuring in forward linked activities. These gains in foreign exchange are relevant when the exporting country suffers from a structural balance of payments deficit which cannot be cured by conventional means. Of the main exporters only Bolivia has serious balance of payments problems and even in this case the country's 1979 devaluation may improve the payments position.

More important is the notion of RV increases and rises in the share of FCV as indicators of increases in real income. This depends on the social opportunity cost of any additional domestic resources used to secure the increases in RV or share in FCV. Where the tin industry is small in relation to the rest of the economy it is possible to consider directly the real income gain generated by the industry as a whole. This was calculated for Malaysia, Thailand and Indonesia up to and including the smelting stage. In general the net economic gain (output value less the social opportunity cost of inputs, expressed as a percentage of the value of output)was much nearer to the figure for gross profit (as a percentage of output) than it was to RV, and RV was generally higher than the gross profit percentage. The NEG coefficients also assumed full local ownership, like most of the investment projects. For dredging therefore a deduction from NEG has to be made to take account of the likely profit outflow. This further raises the divergence between RV and NEG. Two points of caution are necessary in interpreting this result. First, the NEG coefficients were taken from data for individual projects and there may be cost differences between these and the whole of a sector. Second, the investment appraisals of the same projects did show private profitability (before tax) to be less than profitability in terms of net social benefits, and this divergence from private profitability was greater for gravel pumping. However, it is clear that many inputs into tin mining cannot be given a zero shadow price. Petroleum products must be imported at world prices into Thailand and can be sold at world prices from Malaysia and Indonesia. Labour-intensity in electricity is low and its other inputs are in principle tradeable and even labour-intensive engineering is highly using of scarce labour skills. The relatively high labour-intensity of gravel pumping also uses much skilled labour, though one could argue that in the long run a growing demand for skilled labour tends to improve the quality of the workforce - so it is a benefit not a cost. The figures for gross profits, RV and NEG can be summarised as follows:

(PERCENTAGE ANNUAL VALUE OF OUTPUT)

| | GROSS PROFIT | RV | NEG | NEG AFTER PROFIT OUTFLOW |
|---|---|---|---|---|
| MALAYSIA | | | | |
| Dredging | 65.9 | 76.8 | 60.7 | 52.6 |
| Gravel Pumping | 43.3 | 86.0 | 61.0 | 61.0* |
| | | | 41.7 | 41.7** |
| THAILAND | | | | |
| Offshore dredging | 68.7 | 58.8 | n/a | n/a |
| Onshore dredging | 63.1 | 86.3 | 34.6 | 30.2 |
| Gravel pumping | 54.5 | 79.1 | 51.0 | 51.0 |
| Suction boats | 54.7 | n/a | n/a | n/a |
| INDONESIA | | | | |
| Offshore dredging | 61.9 | 93.3 | 56.7 | 56.7 |
| Onshore dredging | 44.3 | 90.9 | n/a | n/a |
| Gravel pumping | 10.3 | 83.8 | 15.9 | 15.9 |

* (large mines)
** (small mines)

Only in the gravel pump mines of Indonesia are economic gains really low and widely divergent from RV. It is possible that the low profitability which this represents is the result of the Indonesian data being for lower grade operations on Belitung, and some of the Bangka gravel pumping mines may well be more profitable. The onshore dredging figures for Thailand probably represent a divergence between the project costs and the sectoral costs which is greater than in the other sectors. It must be remembered that the Malaysian results are achieved with lower grades of ore than in the other producers generally.

In Thailand and Indonesia tin is of minor importance in relation to export earnings, GDP and total employment. In Malaysia it is somewhat more important but it has still been treated here as being small in relation to the rest of the economy. In Bolivia, however, it generates two-thirds of export earnings and clearly the exchange rate in particular would be much different in the absence of tin exports. Thus any potential identification of RV with real income gains would have to take account of the consequent output changes in other sectors (i.e. other exports and import competing industries) in a 'without tin' situation.

The Harvard team estimates that approximately one-third of Bolivian export earnings could be considered as a real income gain if these ramifications were taken into account.

Besides real income gains from raising RV and the local share of FCV, with a given price of tin, attempts can be made to raise real income by raising price, since the demand for tin appears to be inelastic at least in the medium term. This study has tended to dismiss the cartelisation potential of the industry. In spite of many conditions being favourable to cartelisation (a small number of large LDC producers, inelastic demand, inability of new producers greatly to raise supply), the market could be dominated by sales from the US strategic stockpile, currently equivalent to about a year's world output. One other price raising possibility mentioned here has been that export taxation, by reducing supply from Bolivia and marginal gravel pump mines in Asia, may have shifted the incidence of these taxes on to consumers to a limited extent.

The International Tin Agreement has performed a useful function in supporting the floor price through export control. This is extremely helpful to producers in planning investment. It has had little ability to hold the ceiling prices and the US stockpile undoubtedly has been a stronger influence. The current (fifth) agreement expires in 1981 and at the present time negotiations have stalled owing to the USA's insistence that the export control should be abandoned and that the agreement should rely simply on a larger buffer stock.

## 7.3 NEW INVESTMENT AND THE EFFECTS OF POLICY

A substantial part of this research has been concerned with assessing policy towards new investment in tin in the three main intra-marginal producing countries. Investment appraisals have been conducted of dredging and gravel pumping projects both from a private financial point of view and from that of economic and social profitability using the Squire-van der Tak (1975) methodology developed for the World Bank. 'Economic' appraisals using the 'ST' methodology are essentially the same as the well known Little-Mirrlees method of social cost benefit analysis. The 'ST' social appraisals, however, modify the shadow prices to take account of the government's policy towards redistributing income. These appraisals allow a more informed view to be taken of the effectiveness of tax policy and equity sharing in securing gains from foreign enterprises and also of the claims voiced (particularly in Malaysia) that 'penal' taxation is discouraging investment. In addition, they allow some comparison of the technical efficiency of mining in the various countries, of the economic and social profitability of alternative techniques and of the level of price necessary to induce investment. The project has not attempted to make its own price projections, especially in view of the highly inaccurate nature of most published price projections on tin in the past. It is worth noting in this connection, however, that the US Bureau of Mines projects a steady rise in world tin consumption of about 1 per cent per annum to the year 2000, well within know reserve figures. The World Bank predicts a rise in western consumption of 1.5 per cent p.a. over the 1980's and production increases of 1.2 per cent p.a. so present high prices are expected by the Bank to continue (4) though obviously price depends too on future US stockpile release policy; and continuing recession in western countries also would reduce price.

In gravel pumping a comparison was made between very small scale mines (of 20-30,000 cubic metres per month throughput) and, in Malaysia, a larger mine of more than double that size. In Malaysia gravel pumping grades of ground were taken to be approximately 0.12 kg of metal per cubic metre, and double that in the other two countries. The Malaysian mines were also deeper and required considerable dry stripping. Appraisals were also made of large capacity dredges in Malaysia (onshore) and Indonesia (offshore) and a small secondhand dredge in Thailand since data on new dredges there (and there was only one built in the 1970's) could not be obtained. Grades were taken as approximately 0.12 kg in Malaysia and Thailand and 0.24 in Indonesia. It proved impossible to obtain new data on suction boat operations but a study by a local university team in 1976 (Chulalongkorn University, 1976) indicated that at that time their rates of return (at least over a short period) were often in the region of several hundred per cent. They were, of course, working exceptionally high grades of ore.

Analysis was conducted in constant price terms at discount rates of 5 per cent, 10 per cent and 15 per cent at alternative projected tin prices of M$1500, 1700, 2000 and 2250 a picul (5). Costs referred to 1977-8, and 1979 in the case of the Indonesian gravel pump mine. The results can be summarised as follows:

### 7.3.1 Before tax profitability and technical efficiency

In Malaysia the larger GP mine was superior to the smaller at all prices and discount rates from 5 per cent to 15 per cent in terms of net present value (adjusted for output) and also in terms of internal rate of return. This reflects the economies of scale in the larger operation, but many Malaysian deposits are small so that the small mine results are relevant to the situation of many operators. Running the Thailand GP mine at the Malaysian ore grade produced only a 4 per cent IRR at M$2000. Note, however, that a lower grade mine might be run more quickly (i.e. a larger throughput would be used) than a richer mine of the same size, so the technical efficiency comparison is by no means perfect. The Indonesian operation was substantially less profitable than the Thai one at low prices, but was much more price sensitive in its profitability since its initial capital costs were very low. The Indonesian dredge was slightly less profitable than the Malaysian, in spite of working on double the grade. This was mainly a result of its capital costs being more than double the Malaysian. This partly a reflection of the greater technical complexity of an offshore dredge, but probably also that Malaysia has a local and efficient dredge construction industry whereas Indonesia would generally import a complete dredge from abroad.

### 7.3.2 Effects of taxation

Over the 1977-8 range of prices the then Malaysian tax left a large private dredging investment an IRR of 9-12 per cent in real terms, and the larger GP mine with 18 per cent even at $1500. The small mine, however, was sharply hit by taxation and yielded a zero return at $1700, and 9 per cent at $2000. The Thai GP mine earned 11 per cent at $1700, but it would not have been viable at the Malaysian grade. The Indonesian GP mine was similar in profitability to the Thai mine at $1700, but was much more price sensitive, yielding 36 per cent at $2000.

### 7.3.3 Economic profitability and choice of technique

Gravel pumping and dredging inland are competing techniques in the sense that many dredging properties could in principle be worked by multiple gravel pump operations. Moreover, their relative profitability is price sensitive in that dredging often yields higher IRR's and NPV's at low prices and gravel pumping becomes more profitable at higher prices. In fact in these appraisals the IRR's and the output-adjusted NPV's at discount rates of about 10 per cent showed a superiority of the larger gravel pump mine in Malaysia over a large dredge, though more detailed experimental work would be necessary to establish whether and how much costs would rise if a really large multiple GP operation were mounted. In Thailand the secondhand dredge would have been superior to the GP mine at a similar grade, which no doubt is the reason why many former GP miners in Thailand have moved into small dredging. In Indonesia there has been a lack of recent inland dredge building to compare with gravel pumping.

Reappraisal of the results in terms of economic accounting prices tended to favour gravel pumps over new dredges, and in all cases economic profitability was higher than private profitability before tax and export duty. The divergence between economic and private profitability became far less pronounced at high prices since all of the reappraisal was in terms of input prices, marginal export revenue for individual projects being assumed equal to price. This assumption is justified for each producing country as a whole, given that the probable slight expected upward trend in consumption generates a need for increased output at a constant price. The divergence between economic and private profitability rose slightly with the discount rate.

### 7.3.4 Rates of resource depletion

It can be shown in principle (see Appendix) that higher expected prices will lead to a more rapid rate of depletion. This is in the sense that NPV will be higher if more capital equipment is installed to produce a larger annual output, even if that equipment is scrapped before the end of its physical life. In gravel pumping, properties are often small and depleted rapidly anyway and much of the equipment is of short life. To test the relevance for individual dredging projects, a Malaysian project was run with only a five year mining life, but with two dredges installed (at some saving in unit capital costs). The private profitability of this 'half-life' project relative to the single dredge alternative rose with price, but less so in the after tax and export duty version. Before tax and duty, a price of M$2500 and a discount rate of around 10 per cent or more was required for an NPV switch, and in an economic appraisal the NPV switch occurred at a 10 per cent discount rate of M$2250.

### 7.3.5 State equity participation

It is well established that if the state takes equity participation in foreign ventures, its rate of return will tend to fall as its stake rises, because its return includes tax revenue which it gets anyway without contributing equity. Taking again a Malaysian project (since fuller data were available in Malaysia) the 'economic' return to the government was calculated in terms of IRR's and NPV's. As expected,

IRR's fell monotonically with rises in the government's equity share at all projected prices. A government able to borrow money and faced with limited investment outlets in mining might well wish to use NPV as a criterion, however. Here the analysis provides some rather interesting results, since the share of government equity which maximises the government's NPV (for a given project life) varies both with expected price and with the discount rate. For example, at prices of M$1500 and M$1700 at 15 per cent discount rates the government's NPV rises monotonically with the degree of foreign ownership. At M$1500 at 10 per cent and 5 per cent it <u>falls</u> monotonically. At M$1700 the beginnings of a third feature are apparent, which is that the NPV's fall at 10 per cent and 5 per cent discount rates with the increase in foreign share to 75 per cent but rise again slightly by the 100 per cent foreign version but at all prices the NPV's at 5 per cent and 10 per cent discount rates are less with complete foreign ownership than with complete state ownership.

### 7.3.6 Income distributional effects

Reappraisal in ST 'social' terms produced disappointing results, though owing to the poorer quality of the Indonesian data, the social reappraisal was conducted only for Malaysia and Thailand. For almost all projects the ratio of 'economic' to 'social' NPV's and IRR's was well in excess of 0.9 and over 0.95 for most Malaysian projects. To some extent this is a reflection of the relatively low share of shadow wages in output value, and of the fact that data did not permit social reappraisal of indirect employment effects. It is also due to the fact that in both countries mining wages are well above the level of shadow wages so that workers' increased consumption as a result of being employed in mining is given a low social value. In other words although the economic profitability of tin mining, especially gravel pumping, is significantly higher than its private profitability, it would appear in each country to add little more benefit to the nation if its effects on income distribution are taken into account. Whether a transfer of a worker from (say) subsistence agriculture which gives him a substantial non-marginal rise in consumption relative to workers elsewhere is regarded as an insignificant contribution to improved income distribution perhaps is a dubious value judgement, though in large mineral sectors it can contribute to an unbalanced, dualistic economic structure.

## 7.4 THE CHANGING ROLE OF MULTINATIONAL COMPANIES

Recent policies which have increased the host-countries' share of benefits have to some extent been at the expense of accommodation with foreign companies of greater market power than in the past. Charter Consolidated, the foreign partner in the Malaysia Mining Corporation, has interests in more than a dozen metals, whereas the old London Tin Corporation was specialised almost completely on tin mining. Billiton in Thailand, and now again in Indonesia, has become a subsidiary of Royal Dutch Shell, reflecting the tendency of oil companies to diversify into minerals. Rio Tinto Zinc, the largest British mining company has also been a partner in several Malaysian joint ventures. The lack of vertical integration compared (say) to bauxite, nickel or copper has only broken down to a limited degree, however, Billiton in Thailand smelts and markets almost the country's whole output, but in Indonesia smelting and marketing is entirely under government control. Patiño's vast marketing and smelting interests continued after the company's nationalisation in Bolivia, but in 1978 they were sold to the German mining group, Preussag, who have no other interests in tin; and the Malaysia Mining Corporation has been putting pressure on the local smelters to allow the government a stake. Still, tin users have virtually no mining interests, or vice versa and this limits the scope for losses through transfer pricing. The fact that even large tin mining investment projects are small by world mining standards and that efficient small-scale techniques also exist, also has undoubtedly facilitated greater host-country participation. The market power of producer countries has also been strengthened by the fact that they smelt almost the whole of their output.

## 7.5 POLICY IMPLICATIONS

### 7.5.1 Implications for tin exporting countries

1 The gravel pumping sectors in Thailand and Malaysia deserve more encouragement than they have received in the past. They are significantly more labour-intensive than dredging and generate somewhat more favourable linkages. More large properties could be operated by gravel pumps than at present, and the possibility of a multiple gravel pump operation should always be considered as a possible alternative to a dredging investment. High prices (which are likely to continue)(6) increase gravel pumping's relative private and social profitability. The present heavy reliance on export taxes bears down hard on gravel pump mines, especially small ones, when prices are low. When prices are high the after tax profitability of investment in gravel pumping relative to dredging rises sharply and the government's share of this windfall should be greater. The moves in 1979 in Malaysia to relate the export duty threshhold to gravel pump mine costs and to review it regularly are welcome, as is the move to make the duty more progressive. Thailand would be well advised to follow suit, but more reliance on profits tax (as far as problems of compliance would allow) would lead to greater equity in the treatment of different mines.

2 The investment appraisals of dredging in Malaysia suggest that the industry is not too heavily taxed, in spite of its claims to the contrary, and the same is true of Thailand's offshore dredges. The reliance on export taxation does increase the risk of dredging investment, however, and more reliance on profits taxation would be preferable.

3 Suction boats in Thailand clearly need more careful regulation to stop them trespassing on the leases of dredging companies, but otherwise they deserve encouragement. Though their recovery rates are lower than offshore dredges, the tin they are said to 'waste' could easily be won if the area were reworked by other means at some later date. The situation is not like underground mining where low grade ore, once bypassed, may be lost for ever. In terms of employment generation and sheer profitability, the boats are highly beneficial to the Thai economy. They also share the benefits of mineral rents rather more evenly than other methods,

and the Thai government (like the other producers) is committed to achieving income redistribution as well as economic growth.

4 In Indonesia the tin industry has been closed to foreign investment in recent years and is looking to offshore deep dredging for much of its planned output expansion. Deep offshore mining is one area where technical complexity has increased considerably, particularly in the use of computerised techniques, and more foreign investment could be a useful complement to P. T. Timah's own technical expertise in this area.

5 The benefits to the government of different degrees of equity sharing in mining ventures has been shown to be sensitive both to the expected tin price and to the discount rate. A more flexible approach should be adopted therefore to the percentage shares the government opt to take up.

6 Producers should not agree to the dismantling of the export control mechanism in the International Tin Agreement, even if this means the USA will not be a party to the sixth ITA.

7.5.2 Implications for other commodity exporters

1 The same degree of ownership restructuring may differ greatly in its effect on foreign investment according to the manner in which it is carried out. Nationalisation of the Temco lease in Thailand, though modified and followed by similar arrangements to those of Malaysia, had a far sharper adverse effect on foreign business confidence than Malaysia's stock exchange takeover of the London Tin Corporation. The fall in output in Malaysia after the takeover was more due to lease difficulties resulting from federal states conflicts than to restructuring *per se* or taxation.

2 The experience of Bolivia suggests that too strenuous attempts at reallocating resources away from mining can be counter-productive.

3 The experience of the Malaysia Mining Corporation suggests that multinational mining companies may be willing to accept low equity shares and apparently high taxation if the general investment climate is favourable. In these circumstances mining companies can still play a very useful role in providing additional technical expertise, capital and management.

7.5.3 Implications for aid-giving agencies

1 Aid and loan agencies often require the participation of multinational mining companies in projects the agencies help to finance. Although some MNC's are still found in the tin industry, tin is one example where foreign investors' various contributions have been 'depackaged' quite successfully.

2 If the role of MNC's is reduced at the wish of host governments and if MNC's continue to switch their investments to developed mineral producers such as Australia and Canada, help in exploration for new ore deposits is one of the most useful forms of aid for LDC's mineral exporters.

3 The tin industry suggests that a search for more labour-intensive mining methods is not hopeless. If exploration reveals more medium size deposits than giant ones, the increasing tendency to high capital intensity may be slowed down at least, and aid agencies should encourage the development of local mining companies to exploit such deposits.

NOTES

(1) No originality is claimed for these points on Bolivia. They are clearly set out in the Harvard Institute for International Development team's report. (Gillis, 1978, especially chapters seven and eight).
(2) This is very close to the rate estimated by the Harvard team to be necessary to raise the same revenue as the combined export tax and royalty though they recommend a 'Bona Fide Presumptive Tax' (a tax on output with effective allowances for actual differences in costs) as a short term alternative (together with a low production royalty) since they thought that in the late 1970's the Bolivian administrative machinery was not yet adequate to operate a proper income tax (Gillis, 1978, pp.354-59).
(3) The import content of backward linkage production has to be taken into account however.
(4) As reported in TI, June 1980.
(5) The average Penang tin price in 1978 was M\$1743, M\$1961 in 1979 and over the period January-August 1980 it ranged from M\$2088 to M\$2471. In August 1980 it averaged M\$2151.
(6) As this book finally goes to press, in December 1980, the tin price has fallen significantly. This is essentially a feature of the current economic recession in industrial countries. When recovery gets underway, the trend towards high tin prices is likely to reassert itself.

# Appendix

The effect of price changes on the rate of depletion*

The aim of this exercise is to maximise the net present value (Z) of an ore body of size B tons. Annual output (q tons) can be increased but only by shortening the period (n) over which the ore body is depleted. Thus n is also the life of the investment project, and $n = \frac{B}{q}$. It is assumed that price (p) is constant throughout the project and that unit operating costs (c) are also constant over time and with respect to q (and therefore equal to marginal costs). With a given p and c, and B, it is possible to maximise Z with respect to n. The sensitivity of the value of n which maximises Z to alternative levels of p then can be examined. The rate of discount (r) is held constant throughout. The capital cost (A) of the project, which is all incurred in year 0, is assumed to increase in proportion (E) to output, so A = Eq. The capital equipment is assumed to have no scrap value at the end of the project.

Using continuous time, Z can be expressed as follows:

$$Z = \int_0^n (p-c)qe^{-rt}dt - A \qquad \text{Thus}$$

(1) $$Z = \frac{(p-c)\ q\ (1-e^{-rn})}{r} - A$$

(2) $$= \frac{(p-c)}{r}\left(\frac{B}{n}\right)(1-e^{-rn}) - \frac{EB}{n}$$

Maximising Z with respect to project life n:-

Consider $$\frac{dZ}{dn} = \frac{B}{r}(p-c)\frac{d}{dn}\left[\frac{1-e^{-rn}}{n}\right] - EB\frac{d}{dn}\left[\frac{1}{n}\right]$$

(3) $$= \frac{B}{n^2}\left[\frac{p-c}{r}(nre^{-rn} - 1+e^{-rn}) + E\right]$$

When $\frac{dz}{dn} = 0$ then, presuming $B \neq 0$,

(4) $$\frac{p-c}{r}(nre^{-rn} - 1+e^{-rn}) + E = 0$$

(5) $$(nr + 1)e^{-rn} = 1 - \frac{Er}{p-c}$$

It can be shown that (5) has only one positive solution for n (assuming Z to be positive) and that for that positive n $\frac{d^2Z}{dn^2} < 0$, hence it is a maximum (1).

Assuming again that Net Present Value is positive (i.e. that $p-c \geq Er$) we can sketch the curve of $\frac{dZ}{dn}$.

*Written with Dr. R. D. Lee, Lecturer in Mathematics at the University of Essex.

The rough shape is as follows:-

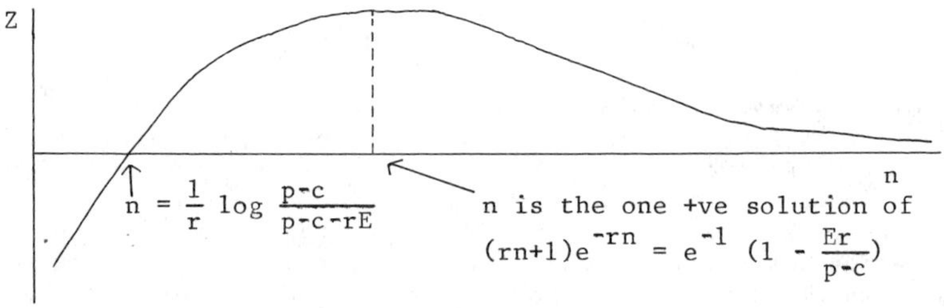


Note that as $n \to \infty$, $Z \to 0$, and that the curve crosses the horizontal axis only once.

<u>As p (price) increases, then n (project life) decreases.</u>

At the turning point $\frac{dZ}{dn} = 0$

$$n = \frac{1}{r} \log \frac{p-c}{p-c-Er}$$

Let p-c=m. As p increases so m increases but $\frac{p-c}{p-c-Er} = \frac{m}{m-Er} = \frac{1}{1-\frac{Er}{m}}$

and as m increases, $\frac{1}{1-\frac{Er}{m}}$ decreases (assuming $1-\frac{Er}{m} > 0$, i.e. $1-\frac{Er}{p-c} > 0$

i.e. NPV >0)

<u>Hence</u> $\frac{p-c}{p-c-Er}$ decreases as p increases

<u>Hence</u> $n = \frac{1}{r} \log \frac{p-c}{p-c-Er}$ decreases as p increases.

In other words, the higher the expected price the more rapidly should the ore body be depleted in order to maximise its Net Present Value.

NOTES

(1) For brevity, the workings in support of this are omitted.

# Bibliography


Allen, G. C. and Donnithorne, A. G., Western Enterprise in Indonesia and Malaya. A Study in Economic Development, Allen and Unwin, London 1954.

Anstee, M. J., Gate of the Sun. A Prospect of Bolivia, Longman, London 1970.

Arief, S., 'The Indonesian Petroleum Industry: A Study of Its Impact on the Indonesian Economy', University of Hull PhD thesis, 1979.

Baldwin, R. E., 'Patterns of Development in Newly Settled Regions', Manchester School, May 1956.

Baldwin, R. E., 'Export Technology and Development from a Subsistence Level', Economic Journal, March 1963.

Barratt Brown, M., The Economics of Imperialism, Penguin, London 1974.

Barry, T. B., 'The Competitive Challenge to Tin', Tin International, November 1978.

Bauer, P. T., 'International Economic Development', Economic Journal, March 1959.

Billiton company, About Billiton, The Hague 1977.

Bird, P. J. W. N., 'The Price Elasticity of Supply of Tin, 1961-75', Malayan Economic Review, October 1978.

Bos, H. C., Saunders, M., and Secchi, C., Private Foreign Investment in Developing Countries. A Quantitative Study on the Evaluation of the Macro-Economic Effects, Reidel Publishing Company, Dordrecht 1974.

Bosson, R. and Varon, B., The Mining Industry and the Developing Countries, Oxford University Press, London 1977.

ter Braake, A. L., Mining in the Netherlands East Indies, Institute of Pacific Relations, New York 1944.

Bruce, C., 'Social Cost-Benefit Analysis: A Guide for Country and Project Economists to the Derivation and Application of Economic and Social Accounting Prices', World Bank Staff Working Paper No. 239, Washington DC 1976.

Caves, R. E., 'International Corporations: The Industrial Economics of Foreign Investment', Economica, February 1971.

Chhabra, J., Grilli, E. and Pollak, p., 'The World Tin Economy: An Econometric Analysis', World Bank Staff Commodity Paper No. 1, Washington DC, June 1978.

Chulalongkorn University 'Report of Research on Small Boat Mining', Mining Engineering Faculty, Bangkok, mimeographed (in Thai) 1976.

CIS, The Rio Tinto Zinc Corporation Ltd. Anti-Report, Counter Information Services, London 1972.

Corden, W. M., 'Exchange Rate Protection', mimeographed, London 1978.

Cullen, P. W. and McDonald, G. C., 'Problems Resulting from the Development of a Large Capacity Mineral Recovery Dredge', Fourth World Tin Conference, mimeographed, Kuala Lumpur 1974.

Department of Mines and Energy, Indonesia, Indonesian Mining Yearbook 1977, Jakarta 1977.

Department of Mineral Resources, Thailand, Tin Production of Thailand: Cost Data Sheet No. 4, Bangkok, April 1978.

Development Consult Inc. PT, Technical and Economic Feasibility Study of Two Offshore Tin Dredgers for PT Tambang Timah in Bangka and Belitung, Volumes A, B and C, Jakarta 1974.

Dimanchi, U. G., Routh, G., and Ali Taha, A. R. E. (eds), Development Paths in Africa and China, Macmillan, London.
Donkers, J. M. and Hauw, N. T., 'Design Aspects of Large Bucket Dredgers', Mining and Transport Engineering, Amsterdam 1975.
Economic and Social Commission for Asia and the Pacific, An Overview of Case Studies on Transnational Corporations in Primary Export Commodities in the ESCAP Region, Joint CTC/ESCAP Unit on Transnational Corporations, Working Paper No. 9, United Nations, Bangkok, October 1979.
Elliott, D., Thailand: Origins of Military Rule, Zed Press, London 1978.
Engel, B. C., and Allen, H. W., Tin Production and Investment, Internanational Tin Council, London 1979.
Evans, D., 'International Commodity Policy: UNCTAD and NIEO in Search of a Rationale', World Development, No. 3 1979.
Fermor, Sir L. L., Report upon the Mining Industry of Malaya, Government Printer, Kuala Lumpur 1939.
Fox, W., Tin, The Working of a Commodity Agreement, Mining Journal Books, London 1974.
Frank, A. G., 'The Development of Underdevelopment', Monthly Review, September 1966.
Fryer, D. W. and Jackson, J. C., Indonesia, Ernest Benn, London 1977.
Gaffnay, M. (ed), Extractive Resources and Taxation, University of Wisconsin Press, Madison, Milwaukee 1967.
Garnaut, R. and Clunies Ross, A., 'Uncertainty, Risk Aversion and the Taxing of Natural Resource Projects', Economic Journal, June 1975.
Geddes, C. F., Patiño, the Tin King, Robert Hale, London 1972.
Gillis, M. with Bucovetsky, Jenkins, G. P., Peterson, U., Wells, L. T. and Wright, B. D., Taxation and Mining: Nonfuel Minerals in Bolivia and Other Countries, Ballinger, Cambridge, Massachusetts 1978.
Gillis, M. and McClure, C. E., 'Incidence of World Taxes on Natural Resources with Special Reference to Bauxite', American Economic Review, Papers and Proceedings, May 1975.
Girvan, N., 'Multinational Corporations and Dependent Underdevelopment in Mineral Export Economies', Social and Economic Studies, 4, 1970.
Gray, L. C., 'Rent under the Assumption of Exhaustibility', Quarterly Journal of Economics, May 1914; reprinted in Gaffnay (1967) to which page numbers refer.
Gregory, R. G., 'Some Implications of the Growth of the Mineral Sector', Australian Journal of Agricultural Economics, August 1976.
Hallwood, C. P., 'Instability in Primary Commodity Markets', Resources Policy, September 1977.
Hallwood, C. P., 'The Profitability of the Buffer Stocks operated under the International Tin Agreements, 1956-77', Resources Policy, December 1979.
Harris, K. L., Tin, Mineral Commodity Profiles MCP-16, US Bureau of Mines, Washington, DC 1978.
Helliwell, J. F., 'Effects of Taxes and Royalties on Copper Mining Investment in British Columbia', Resources Policy, March 1978.
Herfindahl, O. C., 'Some Fundamentals of Mineral Economics', Land Economics, May 1955.
Hicks, J. R., Essays in World Economics, Oxford University Press, London 1959.
Hirschman, A. O., The Strategy of Economic Development, Yale University Press, New Haven 1958.
Hirschman, A. O., 'A Generalised Linkage Approach to Development, with Special Reference to Staples', Economic Development and Cultural Change, Supplement 1977.

Hotelling, H., 'The Economics of Exhaustible Resources', Journal of Political Economy, April 1931.
Hughes, H., 'Economic Rents, the Distribution of Gains from Mineral Exploitation, and Mineral Development Policy', World Development, 11 and 12, 1975.
Hughes, H., and Singh, S., 'Economic Rent: Incidence in Selected Metals and Minerals', Resources Policy, June 1978.
Huntings Technical Services Ltd., South Thailand Regional Planning Study - Sector Studies No. 11, Mining, Songkla, May 1974, published by Overseas Development Administration, London.
Ingram, J. C., Economic Change in Thailand, 1850-1970, Stanford University Press, Stanford, California 1971.
Innis, H. A., The Fur Trade in Canada. An Introduction to Canadian Economic History,1930; revised edition by Clark, S. D. and Easterbrook, W. T., University of Toronto, 1962.
Innis, H. A., Essays in Canadian Economic History (ed Innis, M. Q.), University of Toronto Press, 1956.
International Labour Office, Yearbook of Labour Statistics, Geneva 1979.
International Tin Council, The International Implications of United States Disposal of Stockpiled Tin, London 1973.
International Tin Council, Aspects of the Marketing of Tin, London 1974.
Jackson, J. C., 'Mining in 18th Century Bangka: the Pre-European Exploitation of a "Tin Island"', Pacific Viewpoint, September 1969.
Jones, L. P.,. 'The Measurement of Hirschmanian Linkages', Quarterly Journal of Economics, May 1976.
Jones, W. R., Tinfields of the World, Mining Publications, London 1925.
Khan, K. R., 'The International Tin Agreement 1976: An Assessment of Its Regulatory Machinery As An Instrument of International Policy', Resources Policy, June 1979.
Killick, T., 'The Benefits of Foreign Direct Investment and Its Alternatives: An Empirical Investigation', Journal of Development Studies, January 1973.
Klein, H. S., 'The Creation of the Patiño Tin Empire', Inter-American Economic Affairs, Autumn 1965.
Knights, I. A., 'Deep Water Tin Mining', Tin International, August 1980.
Lall, S., 'Transnationals, Domestic Enterprises, and Industrial Structure in Host LDC's: A Survey', Oxford Economic Papers, July 1978.
Lall, S., and Streeten, P., Foreign Investment, Transnationals and Developing Countries, London Macmillan Press 1977.
Lam, N. V., 'Incidence of Tin Export Taxation in West Malaysia', The Developing Economies, December 1978.
Lanning, G. and Mueller, M., Africa Undermined: Mining Companies and the Underdevelopment of Africa, Penguin, London 1979.
Lecomber, R., The Economics of Natural Resources, Macmillan, London 1979.
Lim, D., 'Industrial Processing and Location: A Study of Tin', World Development, vol. 8, 1980.
Lim Chong Yah, Economic Development of Modern Malaya, Oxford University Press, Kuala Lumpur 1967.
Lipton, M., Why Poor People Stay Poor. A Study of Urban Bias in World Development, Temple Smith, London 1977.
Little, I. M. D., and Mirrlees, J. A., Manual of Industrial Project Analysis in Developing Countries (in two volumes), Organisation for Economic Cooperation and Development, Paris 1968.
Little, I. M. D., and Mirrlees, J. A., Project Appraisal and Planning for Developing Countries, Heinemann, London 1974.
Lloyd, B. and Wheeler, E., 'Brazil's Mineral Development: Potential and Problems', Resources Policy, March 1977.

Lloyds Bank, Economic Report on Bolivia, Overseas Division, London, 1980.
Lo Sum Yee, The Development Performance of West Malaysia, Heinemann, Kuala Lumpur 1972.
Mackie, J. A. C., 'The Indonesian economy, 1950-63', in Glassburner, B. (ed), The Economy of Indonesia, Selected Readings, Cornell University Press, Ithaca 1971.
Mantell, C. L., Tin. Its Mining, Production, Technology and Applications, Hafner Publishing, New York 1970 (reprint of 1949 edition).
McKern, R. B., Multinational Enterprise and Natural Resources, McGraw-Hill, Sydney 1976.
Meier, G. M., 'International Trade and International Inequality', Oxford Economic Papers, October 1958.
Meier, G. M., The International Economics of Development, Harper, New York 1968.
Merrett, A. J. and Sykes, J., The Finance and Analysis of Capital Projects, Longmans, London 1973.
Mikesell, R. F. (ed), Foreign Investment in the Petroleum and Mineral Industries. Case Studies in Investor-Host Country Relations, John Hopkins, Baltimore 1971.
Mikesell, R. F., Foreign Investment in Copper Mining. Case Studies of Mines in Peru and Papua New Guinea, John Hopkins, Baltimore 1975.
Mikesell, R. F., New Patterns of World Mineral Development, British North American Committee, London 1979.
Mollema, J. C., De Ontwikkeling van het Eiland Billiton en van de Billiton - Maatschappij, Nijhoff, The Hague 1922.
Murray, A. M., 'A Tin Exchange in Malaysia', paper presented to National Seminar on the Mining Industry, Kuala Lumpur, August 1977.
Myint, H., 'An Interpretation of Economic Backwardness', Oxford Economic Papers, June 1954.
Myint., H., The Economics of the Developing Countries, Hutchinson, 3rd edition, London 1967.
Myrdal, G., Economic Theory and Underdeveloped Regions, Duckworth, London 1957.
Nankani, G., 'Development Problems of Mineral Exporting Countries', World Bank Staff Working Paper No. 354, Washington DC 1979.
Nankani, G., 'Development Problems of Nonfuel Mineral Exporting Countries', Finance and Development, March 1980.
National Union of Mineworkers, Bolivia. Report of an NUM Delegation in July 1979, London 1979.
Pack, H., 'The Employment-Output Trade-Off in LDC's - A Microeconomic Approach', Oxford Economic Papers, 3, 1974.
Page, W., 'Mining and Development: Are They Compatible in South America?', Resources Policy, December 1976.
Page, W., 'One Attempt at Taking a Long Term View to Assist Decision Makers in the World Tin Industry - An Interim Report', Science Policy Research Unit, University of Sussex, mimeographed, 1977.
Pajon Sinlapajan, 'Tin Dredging in Thailand' in International Tin Council, A Second Technical Conference on Tin, Bangkok 1969.
Palmer, I., The Indonesian Economy Since 1965, Cass, London 1978.
Pearson, S. R. and Cownie, J., Commodity Exports and African Economic Development, Heath, Lexington 1974.
Peterson, F. M., and Fisher, A. C., 'The Exploitation of Extractive Resources: A Survey', Economic Journal, December 1977.
Prebisch, R., Towards a Dynamic Development Policy for Latin America, UN, New York 1963.
Putucheary, J. J., Ownership and Control in the Malayan Economy, Eastern Universities Press, Singapore 1960.

Rachan Kanjan-Vanit, Pow Kham-Ourai and Champion, W., 'Offshore Dredging of Tin Deposits in South Thailand', in International Tin Council, A Second Technical Conference on Tin, Bangkok 1969.
Radetzki, M., 'Where Should Developing Countries' Minerals be Processed? The Country View Versus the Multinational Company View', World Development, 1977.
Radetzki, M., 'Market Structure and Bargaining Power. A Study of Three International Mineral Markets', Resources Policy, June 1978.
Raphael, Zorn, Manual of Palm Oil/Rubber and Tin Producing Companies, privately published, London 1976.
Raw, C., Slater Walker. An Investigation of a Financial Phenomenon, Deutsch, London 1977.
Reuber, G. L. (with Crokell, H., Emerson, M., and Gallais-Hamouno, G.), Private Foreign Investment in Development, Clarendon Press, Oxford, for OECD Development Centre 1973.
Richardson, G. B., 'The Organisation of Industry', Economic Journal, September 1972.
Riedel, J., 'A Balanced-Growth Version of the Linkage Hypothesis: A Comment', Quarterly Journal of Economics, May 1976.
Robertson, W., Report on the World Tin Position, with Projections for 1965 and 1970, International Tin Council, London 1964.
Roemer, M., 'Resource-Based Industrialisation in Developing Countries - A Survey', Journal of Development Economics, June 1979.
Sen, A. K., Employment, Technology and Development, Clarendon Press, Oxford 1975.
Singer, H. W., 'The Distribution of Gains between Investing and Borrowing Countries', American Economic Review, Papers and Proceedings, May 1950.
Singer, H. W., 'The Distribution of Gains from Trade and Investment - Revisited, Journal of Development Studies, July 1975.
Singer, H. W., 'Beyond Commodity Policy: Structural Changes and Financial Compensation', Institute of Development Studies Bulletin, Sussex, No. 4, 1976.
Slater, J., Return to Go. My Autobiography, Futura Publications, London, 1978.
Smith, D. N. and Wells, L. T., Negotiating Third-World Mineral Agreements Promises as Prologue, Ballinger, Cambridge, Massachussetts 1975.
Smith, G. W. and Schink, G. R., 'The International Tin Agreement: A Reassessment', Economic Journal, December 1976.
Snape, R. H., 'Effects of Mineral Development on the Economy', Australian Journal of Agricultural Economics, December 1977.
Solow, R. M., 'The Economics of Resources or the Resources of Economics', American Economic Review, May 1974.
Squire, L., and van der Tak, Economic Analysis of Projects, World Bank Research Publications, John Hopkins University Press, Baltimore 1975.
Sriwichchar Bunyapisit and Pithaya Boonying, 'Report on Tin Research Study', Industrial Research Section, Department of Business Economics, Ministry of Commerce, Bangkok (in Thai) 1977.
Steele, H., 'Natural Resource Taxation: Resource Allocation and Distribution Implications', in Gaffnay (1967).
Stewart, F., 'Choice of Technique in Developing Countries', Journal of Development Studies, October 1972.
Stewart, F., Technology and Underdevelopment, 2nd edition, Macmillan, London 1978.
Stigler, G. J., 'The Division of Labour is Limited by the Extent of the Market', Journal of Political Economy, June 1951.

Stiglitz, J., 'Monopoly and the Rate of Extraction of Exhaustible Resources', American Economic Review, September 1976.
Streeten, P. P., Economic Integration. Aspects and Problems, Sythoff, Leyden 1961.
Sumitro Djojohadikusomo, 'Future Perspectives of Indonesia's Economy', Jakarta, mimeographed, 1978.
Thoburn, J. T., 'Commodity Prices and Appropriate Technology: Some Lessons from Tin Mining', Journal of Development Studies, October 1977a.
Thoburn, J. T., Primary Commodity Exports and Economic Development. Theory, Evidence and a Study of Malaysia, Wiley, London 1977b.
Thoburn, J. T., 'High Prices Favour Small Producers', Tin International, February 1978a.
Thoburn, J. T., 'Malaysia's Tin Supply Problems', Resources Policy, March 1978b.
Thoburn, J. T. and Camina, A. R., 'The Influence of Product Price on the Choice of Technique', South East Asian Economic Review, forthcoming.
Thoburn, J. T. and Widyono, B., Transnational Corporations and the Distribution of Gains in the Tin Industry of South East Asia, Economic and Social Commission for Asia and the Pacific, Joint CTC/ESCAP Unit on Transnational Corporations Working Paper No. 7, United Nations, Bangkok, August 1979.
Timah, P. N., Tin In Indonesia, Jakarta 1972.
Timah, P. T., Tin in Indonesia, Jakarta 1976.
United Nations, Yearbook of International Trade Statistics, New York 1978a.
United Nations, Yearbook of National Accounts Statistics, New York 1978b.
UNIDO, World Industry Since 1960: Progress and Prospects, New York 1979.
United States Embassy Indonesia, Industrial Outlook: Minerals, Jakarta, August 1977.
Veitch, M. D., 'National Parameters for Project Appraisal in Malaysia', UNDP/World Bank, (in two volumes) Kuala Lumpur 1977.
Vernon, R., Sovereignty at Bay, The Multinational Spread of US Enterprises, Penguin, London 1973.
Wall, D., 'Industrial Processing of Natural Resources', World Bank Commodity Working Paper No. 4, Washington, DC 1979.
Watkins, M. H., 'A Staple Theory of Economic Growth', Canadian Journal of Economics and Political Science, May 1963.
Watkins, M. H., 'The Staple Theory Revisited', Journal of Canadian Studies, winter 1977.
Weisskopf, R. and Wolff, E., 'Linkages and Leakages: Industrial Tracking in an Enclave Economy', Economic Development and Cultural Change, July 1977.
White, L. J., 'The Evidence on Appropriate Factor Proportions for Manufacturing in Less Developed Countries: A Survey', Economic Development and Cultural Change, October 1978.
Widyono, B., Transnational Corporations in Export-Orientated Primary Commodities: A Study of Relative Bargaining Positions and the Distribution of Gains, Case Study No. 1: The Tin Industry of Bolivia, CEPAL/CTC Joint Unit, 1977.
Williams, T. A. J., 'The Valuation and Appraisal of Alluvial Tin-Ore Deposits', University of London, MSc thesis, 1973.
Wit Satyarakwit, 'Tin: A Comparison of Gravel Pump and Dredging Mining', Thammasat University, MEcon thesis, Bangkok 1971.
Wong, J., ASEAN Economies in Perspective. A Comparative Study of Indonesia, Malaysia, the Phillipines, Singapore and Thailand, Macmillan, London 1979.

Wong Lin Ken, The Malayan Tin Industry to 1914, With Special Reference to the States of Perak, Selangor, Negri Sembilan and Pahang, University of Arizona Press, Tucson 1965.
World Bank, Commodity Trade and Price Trends, Commodities and Export Projections Division, Report No. EC-166/79, Washington, DC 1979.
World Bank, Thailand: Towards a Development Strategy of Full Participation, Washington, DC 1980a.
World Bank, World Development Report, 1980, Oxford University Press, London 1980b.
Yeats, A. J., Trade Barriers Facing Developing Countries: Commercial Policy Measures and Shipping, London, Macmillan 1979.
Yip Yat Hoong, The Development of the Tin Mining Industry of Malaya, Singapore, University of Malaya Press 1969.
Zaalberg, P. H. A., 'Offshore Tin Dredging in Indonesia', Transactions, Section A, Institution of Mining and Metallurgy, London 1970.
Zondag, C. H., The Bolivian Economy, 1952-65. The Revolution and Its Aftermath, Praeger, New York 1966.
Zorn and Leigh-Hunt, Manual of Rubber and Tin Producing Companies, privately published, London 1974.

# Index

References from Notes indicated by 'n' after page reference